T0140382

Advances in Experimental Medicine and Biology

Clinical and Experimental Biomedicine

Volume 1133

Subseries Editor
Mieczyslaw Pokorski

More information about this subseries at http://www.springer.com/series/16003

Mieczyslaw Pokorski

Editor

Advances in Medicine and Medical Research

 Springer

Editor
Mieczyslaw Pokorski
Opole Medical School
Opole, Poland

ISSN 0065-2598   ISSN 2214-8019   (electronic)
Advances in Experimental Medicine and Biology
ISSN 2523-3769   ISSN 2523-3777   (electronic)
Clinical and Experimental Biomedicine
ISBN 978-3-030-12925-5   ISBN 978-3-030-12923-1   (eBook)
https://doi.org/10.1007/978-3-030-12923-1

Library of Congress Control Number: 2019933306

© Springer Nature Switzerland AG 2019
This work is subject to copyright. All rights are reserved by the Publisher, whether the whole or part of the material is concerned, specifically the rights of translation, reprinting, reuse of illustrations, recitation, broadcasting, reproduction on microfilms or in any other physical way, and transmission or information storage and retrieval, electronic adaptation, computer software, or by similar or dissimilar methodology now known or hereafter developed.
The use of general descriptive names, registered names, trademarks, service marks, etc. in this publication does not imply, even in the absence of a specific statement, that such names are exempt from the relevant protective laws and regulations and therefore free for general use.
The publisher, the authors, and the editors are safe to assume that the advice and information in this book are believed to be true and accurate at the date of publication. Neither the publisher nor the authors or the editors give a warranty, express or implied, with respect to the material contained herein or for any errors or omissions that may have been made. The publisher remains neutral with regard to jurisdictional claims in published maps and institutional affiliations.

This Springer imprint is published by the registered company Springer Nature Switzerland AG.
The registered company address is: Gewerbestrasse 11, 6330 Cham, Switzerland

Contents

Adv Exp Med Biol - Clinical and Experimental Biomedicine (2019) 4: 1–8
https://doi.org/10.1007/5584_2018_304
© Springer Nature Switzerland AG 2018
Published online: 4 December 2018

Left Ventricular Strain and Relaxation Are Independently Associated with Renal Cortical Perfusion in Hypertensive Patients

Arkadiusz Lubas, Robert Ryczek, Artur Maliborski, Przemysław Dyrla, Longin Niemczyk, and Stanisław Niemczyk

Abstract

Renal perfusion, which depends on cardiac function, is a factor conditioning the work of kidneys. The objective of the study was to assess the influence of cardiac function, including left ventricular contractility and relaxation, on renal cortical perfusion in patients with hypertension and chronic kidney disease treated pharmacologically. There were 63 patients (7 F and 56 M; aged 56 ± 14) with hypertension and stable chronic kidney disease enrolled into the study. Serum cystatin C, with estimated glomerular filtration rate (eGFR), ambulatory blood pressure monitoring, carotid intima-media thickness (cIMT),

echocardiography with speckle tracking imaging and the calculation of global longitudinal strain (GLS), diameter of vena cava inferior (VCI), and an ultrasound dynamic tissue perfusion measurement of the renal cortex were performed. We found that the renal cortical perfusion correlated significantly with age, renal function, cIMT, GLS, left ventricular ejection fraction (LVEF), left ventricular mass index (LVMI), diastolic peak values of early (E) and late (A) mitral inflow velocities ratio (E/A) and E to early diastolic mitral annular tissue velocity (E/E'), but not with VCI, or the right ventricle echocardiographic parameters. In multivariable regression analysis adjusted to age, only eGFR, E/E', and GLS were independently related to renal cortical perfusion ($r^2 = 0.44$; $p < 0.001$). In conclusion, the intensity of left ventricular strain and relaxation independently influence renal cortical perfusion in hypertensive patients with chronic kidney disease. A reduction in left ventricular global longitudinal strain is superior to left ventricular ejection fraction in the prediction of a decline in renal cortical perfusion.

A. Lubas and S. Niemczyk
Department of Internal Diseases, Nephrology and Dialysis, Military Institute of Medicine, Warsaw, Poland

R. Ryczek
Department of Cardiology and Internal Diseases, Military Institute of Medicine, Warsaw, Poland

A. Maliborski
Department of Radiology, Military Institute of Medicine, Warsaw, Poland

P. Dyrla
Department of Gastroenterology, Military Institute of Medicine, Warsaw, Poland

L. Niemczyk (✉)
Department of Nephrology, Dialysis and Internal Medicine, Warsaw Medical University, Warsaw, Poland
e-mail: lniemczyk@wum.edu.pl

Keywords

Blood pressure · Cardiac function · Doppler ultrasound · Hypertension · Kidney disease · Left ventricular ejection · Left ventricular strain · Renal function

1   Introduction

Renal perfusion depends on cardiac function. Maintaining a proper renal blood circulation is a prerequisite for normal kidney functioning. It is believed that a reduction in the cardiac output to less than 1.5 L/min/1.73m^2 results in deficiency of blood flow and oxygenation of kidneys leading to their functional deterioration (Ljungman et al. 1990). In the physiological condition, intrarenal mechanisms responsible for the renal blood flow autoregulation ensure stabilization of renal function in the mean arterial pressure range of 70–130 mmHg (Burke et al. 2014). However, many drugs used to treat hypertension interfere with the mechanisms of renal autoregulation, which results in the kidney exposure to arterial pressure fluctuations (Meyrier 2015). On the other hand, intrarenal changes occurring in the course of a chronic kidney disease also limit the efficiency of renal autoregulation. Previous studies have shown a significant correlation of renal perfusion, assessed by ultrasound dynamic tissue perfusion measurement, with biochemical (troponin I, NT-proBNP), functional (left ventricular ejection fraction (LVEF), cardiac index (CI), and stroke volume), and structural (left ventricular mass index (LVMI)) indices of cardiac function (Lubas et al. 2013, 2015). In patients with acute decompensated heart failure, there is a greater influence of venous congestion and elevated pressure in the right ventricle than that of reduced LVEF on the deterioration of renal function (Gnanaraj et al. 2013). Further, studies on the cardiotoxicity of anticancer drugs have shown that severe disturbances in the contractility of left ventricular longitudinal fibers, expressed as a global longitudinal strain (GLS), can be determined before the decrease in LVEF becomes significant (Smiseth et al. 2016). Likewise, reduction in GLS with normal LVEF has been reported in patients with three-vessel coronary artery disease, heart failure with preserved LVEF, and tight aortic stenosis (Attias et al. 2013; Choi et al. 2009; Liu et al. 2009). All that suggests a greater sensitivity of GLS, compared to LVEF, in the detection of subclinical damage to the left ventricular muscle.

The objective of the present study was to assess the influence of hemodynamic cardiac function, including contractility of the left ventricular longitudinal fibers, on renal perfusion in pharmacologically treated patients with hypertension and chronic kidney.

2   Methods

The study included 63 consecutive patients (7 F and 56 M; aged 56 ± 14) with hypertension and stable chronic kidney disease, seeking medical attention in the nephrology clinic. Exclusion criteria were acute cardiac or renal disease, glomerulonephritis requiring immunosuppressive therapy, renal focal lesions disallowing the adequate ultrasound assessment of renal perfusion, pelvicalyceal system dilatation, past renal artery stenosis, chronic kidney disease in stage 5 (National Kidney Foundation 2002), symptoms of heart failure, significant valvular heart disease, segmental contractility disturbances noticed in echocardiography, previous myocardial infarct, arrhythmias, tachycardia found in the perfusion ultrasound or echocardiography, hyperkinetic state, active inflammation, cardiorenal diseases in the course of other pathologies (connective tissue diseases, diabetes mellitus, or amyloidosis), generalized neoplastic disease, and current or past anticancer treatment.

2.1   Cardiorenal Function

Ambulatory 24-h blood pressure monitoring was conducted in all patients using an ABPM-04 monitor (Meditech, Budapest, Hungary), with measurements taken every 15 min during daytime and every 30 min during nighttime.

Carotid Sonography To assess the severity of vascular changes occurring in the course of arterial hypertension, which can affect renal perfusion, each patient underwent a measurement of left common carotid artery intima-media thickness (cIMT) using the 11 L transducer

(10–13 MHz). To calculate cIMT, three manual measurements of intima-media complex on the far wall of the left carotid artery, at least 10 mm from the carotid sinus, were averaged.

Cardiac Sonography Echocardiography was performed at the day when the blood pressure monitor was disconnected, using the Vivid S6 system with the M4S-RS sector transducer of 1.5–3.6 MHz (GE Healthcare, Chicago, IL). Measurements of the tricuspid annular plane systolic excursion (TAPSE), diameter of vena cava inferior (VCI), and wall thickness and diameter of the left ventricular cavity were performed in the M-mode imaging in accordance with the recommendations of the American Society of Echocardiography (Sahn et al. 1978). Left ventricular mass was calculated from the Devereux et al. (1986) formula and then normalized for body surface area, calculated according to Mosteller (1987), to obtain the LVMI index. The LVEF in the Simpson's biplane method, CI, and the indices of diastolic LV function such as the peak values of early (E wave) and late (A wave) mitral inflow velocities and tissue Doppler early diastolic mitral annular velocity (E′) at the septal corner of mitral annulus were measured. Then, E/A and E/E′ ratios were calculated as previously described (Lubas et al. 2017; Lang et al. 2006). The LV longitudinal strain calculations were based upon the speckle tracking technique, using the dedicated automated functional imaging protocol. The measurements were done at post-processing, using electrocardiography-gained loop images acquired in typical apical views. The tracked area was detected automatically and then manually corrected. The GLS was the arithmetic mean of longitudinal strain in four-, two-, and three-chamber views.

Kidney Ultrasound After having the biochemistry tests done, patients underwent kidney ultrasound examination using the 4 L 2–5 MHz convex transducer (Logiq P6; GE Healthcare, Seoul, Korea). Renal perfusion was assessed using the same transducer in the color Doppler mode, as previously described (Lubas et al. 2013, 2015). Short sequences of video clips recorded in the DICOM format presenting the color-coded flow velocity in the renal cortex were then evaluated (PixelFlux medical device; Chameleon-Software, Leipzig, Germany) according to the ultrasound dynamic tissue perfusion measurement method (Scholbach et al. 2004). The size of averaged total (arterial and venous) renal cortical perfusion (RCP) in cm/s was evaluated as a surrogate of right and left heart function on kidney perfusion.

Estimated glomerular filtration rate (eGFR) was assessed using the chronic kidney disease epidemiology formula based on the measurement of cystatin C in the serum (Levey et al. 2009).

2.2 Statistical Analysis

Data were presented as means ±SD. Correlations between variables were examined using the Pearson or Spearman method depending on the fulfillment of the normal distribution condition. The presence of intragroup differences was tested using the Kruskal-Wallis analysis of variance (ANOVA). Stepwise multivariable regression analysis was used to identify the factors independently associated with RCP. Receiver operating characteristic (ROC) analysis was performed to identify the nadir value of LVEF corresponding to the threshold value of RCP. For statistical evaluation, a commercial Statistica 12 package was used (StatSoft Inc., Tulsa, Oklahoma, USA).

3 Results

Baseline patient characteristics and the results of echocardiography and RCP measurements are presented in Table 1.

In the study group, hypertension was treated with angiotensin-converting enzyme inhibitors or angiotensin receptor blockers in 36 patients. Thirty-six patients received β-blockers, 23 calcium channel blockers, and 38 diuretics, and 12 received α-blockers. RCP associated

Table 1 Baseline patient characteristics and results

Indices	All patients ($n = 63$)	RCP Correlation coefficient
Age (yr)	55.6 ± 13.9	−0.349*
BMI (kg/m^2)	28.4 ± 3.6	−0.055
Cystatin C (mg/dL)	1.47 ± 0.60	−0.636*
eGFR (ml/min/1.73 m^2)	59.2 ± 30.8	0.663*
SBP (mmHg)	126.2 ± 16.1	−0.205
DBP (mmHg)	75.7 ± 11.2	0.039
MAP (mmHg)	92.6 ± 12.7	−0.047
PP (mmHg)	50.5 ± 10.3	−0.270
cIMT (mm)	0.83 ± 0.22	−0.373*
VCI (cm)	1.77 ± 0.39	0.149
TRPG (mmHg)	22.3 ± 8.3	0.009
TAPSE (cm)	2.4 ± 0.4	0.104
LVMI (g/m^2)	102.7 ± 33.3	−0.300*
LVEF (%)	61.6 ± 9.3	0.299*
CI (L/min/m^2)	4.4 ± 1.2	−0.101
E/A	1.08 ± 0.48	0.330*
E/E′	10.2 ± 2.8	−0.414*
GLS (%)	−17.2 ± 3.9	−0.427*
RCP (cm/s)	0.25 ± 0.18	–

Data are means ±SD and correlation coefficients; *BMI* body mass index, *eGFR* estimated glomerular filtration rate, *CI* cardiac index, *cIMT* carotid intima-media thickness, *E/A* early and late mitral inflow velocities ratio, *E/E′* early mitral inflow velocity and mitral annular early diastolic velocity ratio, *LVEF* left ventricular ejection fraction, *GLS* left ventricular global longitudinal strain, *LVMI* left ventricular mass index, *SBP, DBP, MAP, PP,* systolic, diastolic, mean arterial pressure and pulse pressure, *VCI* diameter of vena cava inferior, *TAPSE* tricuspid annular plane systolic excursion, *TRPG* Bernoulli equation-derived pressure gradient from the peak tricuspid regurgitation velocity, *RCP* renal cortical perfusion
*$p < 0.05$

Table 2 Comparison of left ventricular ejection fraction (LVEF) and left ventricular global longitudinal strain (GLS) on renal cortical perfusion (RCP) in groups stratified by quartiles of RCP

	Quartile I ($n = 16$)	Quartile II ($n = 16$)	Quartile III ($n = 16$)	Quartile IV ($n = 15$)	p-value quartile I:II:III:IV
RCP (cm/s)	0.06 ± 0.04*	0.16 ± 0.30*	0.29 ± 0.05*	0.52 ± 0.13*	<0.001
eGFR (mL/min/1.73 m^2)	34.6 ± 14.1*	48.8 ± 23.6**	74.8 ± 26.7*	81.7 ± 33.2**	<0.001
GLS (%)	−14.6 ± 3.9***	−17.6 ± 3.6	−17.6 ± 3.4	−19.2 ± 3.3***	0.011
E/E′	11.6 ± 3.0	11.2 ± 3.3	9.1 ± 2.0	9.0 ± 1.9	0.015
LVEF (%)	57.1 ± 11.5	62.2 ± 8.2	61.6 ± 8.4	65.9 ± 7.2	0.094

eGFR estimated glomerular filtration rate, *E/E′* early mitral inflow velocity and mitral annular early diastolic velocity ratio
*$p < 0.001$; **$p = 0.033$; ***$p = 0.007$

significantly with age, renal function, LVMI, and with the functional indices of the left ventricle (Table 1). However, the TAPSE and TRPG indices of right ventricular function assessed in the standard echocardiographic examination and the diameter of VCI failed to associate with RCP.

Although the initial LVEF significantly associated with RCP, after dividing patients into four groups by quartiles of RCP, the Kruskal-Wallis ANOVA showed significant differences only in the E/E′, GLS, and renal function, but not in LVEF (Table 2; Fig. 1; Panel I and II). In

Fig. 1 Imaging examples of renal cortex perfusion in a patient with normal left ventricular global longitudinal strain (GLS) (**Panel I**) and with abnormal GLS (**Panel II**)

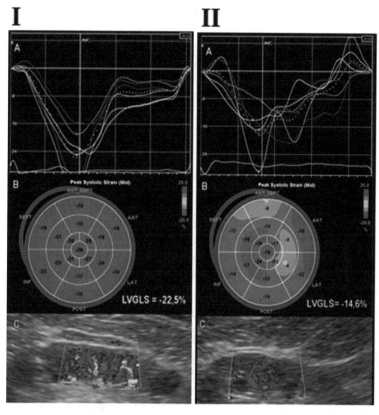

Left ventricular strain (A, B) and renal cortex Doppler perfusion (C) imaging in a 66-year-old man with cystatin C 2.1 mg/dl, E/E' 12.0, LVEF 60.5%, GLS 22.5%, and RCP 0.13 cm/s

Left ventricular strain (A, B) and renal cortex Doppler perfusion (C) imaging in a 66-year-old man with cystatin C 2.0 mg/dl, E/E' 6.5, LVEF 63.5%, GLS 14.6%, and RCP 0.05 cm/s

the multivariable regression analysis performed in relation to age and taking into account the indices that associated with RCP, i.e., LVMI, LVEF, GLS, E/E', IMT, PP, and eGFR, only the eGFR ($r = 0.418$; $p < 0.001$), E/E' ($r = -0.259$, $p = 0.022$), and GLS ($r = -0.238$, $p = 0.035$) were found to independently affect the value of RCP ($r^2 = 0.440$, $p < 0.001$; power of the test 0.999 for probability of type I error $\alpha = 0.05$). The examples of predicted RCP values with known LV GLS, E/E', and eGFR values are shown in Table 3.

The use of the known threshold values for the results of normal and abnormal indices that significantly related to RCP enabled to predict the threshold value of RCP = 0.284 (95%CI 0.231; 0.337). In order to obtain an approximate LVEF result corresponding to the threshold value of RCP, ROC analysis was performed in which the highest sensitivity (70%) and specificity (52%) were found for LVEF of 64.8%. However, this result was insignificant (AUC = 0.611, $p = 0.121$).

4   Discussion

The present study demonstrates that indices of renal function and left ventricular systolic and diastolic function are independent factors enable to explain approximately 44% of the variability in

Table 3 Prediction of renal cortical perfusion (RCP) values based on multivariate regression analysis

eGFR (mL/min/1.73 m²)	GLS (%)	E/E′	Predicted RCP (95%CI) (cm/s)
90	−16	8	0.36 (0.30; 0.42)
60	−16	8	0.28 (0.23; 0.34)
60	−20	8	0.33 (0.27; 0.39)
60	−16	10	0.25 (0.21; 0.29)

eGFR estimated glomerular filtration rate, *GLS* left ventricular global longitudinal strain, *E/E′* early mitral inflow velocity and mitral annular early diastolic velocity ratio, *95%CI* 95% confidence intervals

RCP. GLS assessed by speckle tracking echocardiography is a measure of the averaged strain of the longitudinal myocardium fibers in all segments of the left ventricle, examined during contraction. The shortening of longitudinal fibers is expressed in negative percentage values, where the higher is the absolute value the greater the shortening of the fibers in question (normal GLS < 16%) (Yingchoncharoen et al. 2013). GLS is considered an equivalent to the left ventricular systolic function, and it appears an earlier and more sensitive marker of left ventricular muscle damage than LVEF (Smiseth et al. 2016). Moreover, GLS has been found of predictive value for LVEF decrease in patients with heart failure (Adamo et al. 2017; Romano et al. 2017). GLS is a quantitative, objective method, mostly dependent on the software, and is characterized by a greater repeatability (intra-observer 3.3%; inter-observer 4.0%), while the variability of LVEF results, largely conditioned by the experience of the experimenter, is nearly twice as high (intra-observer 6.3%, inter-observer 10.3%) (King et al. 2016). In the present study, GLS had a greater impact on the RCP value than the calculated LVEF. Although the attempt to estimate the LVEF value corresponding to the borderline RCP was estimated at approximately 65%, this result was not significant and due to the Ray et al. (2010) criteria, had a poor diagnostic value.

The E/E′ ratio is a quotient of the maximal velocity of early mitral valve inflow (E) and early diastolic velocity of the mitral annulus (E′) and it correlates with the left ventricular filling pressure that is higher in case of left ventricular diastolic dysfunction. In this study, E/E′ was significantly inversely associated with renal perfusion, and it was an independent factor modifying

the RCP value. This means that the higher the E/E′ ratio (normal <8 cm/s, abnormal >13 cm/s), the lower the renal perfusion. Thus, both systolic (GLS) and diastolic (E/E′) left ventricular function affect renal perfusion. The evaluation of the left ventricular function is therefore essential for the diagnosis of disorders of the cardiorenal axis. On the other hand, it may be expected that patients with heart failure with preserved ejection fraction (HFpEF), i.e., so-called isolated diastolic insufficiency, would also be characterized by lower RCP indices compared to those without heart failure with the same LVEF value. In this sense, the evaluation of RCP can be useful in the early diagnosis of cardiorenal syndrome.

The study by Gnanaraj et al. (2013) has shown a significant role of venous congestion in reducing renal perfusion in patients with acute decompensated heart failure, while the observed markers of decreased cardiac output are not of such significance. However, that study has used LVEF, without the GLS assessment. Further, in patients with decompensated heart failure, venous congestion is the main symptomatic problem forcing the patients to seek hospital help. It is thus difficult to expect that the impaired renal venous outflow will not cause a decrease in renal blood flow, particularly in the presence of reduced arterial perfusion. In the present study, patients were in the optimal condition, with a well-controlled arterial pressure and without symptoms of heart failure. The functional indices of the right heart (TAPSE, TRPG) and VCI diameter measured directly before the right atrium were not significantly related to renal perfusion.

The majority of antihypertensive drugs used in the treatment of high blood pressure impair renal autoregulation and significantly modify renal perfusion (Digne-Malcolm et al. 2016). Hence, there is a greater dependence of renal perfusion on functional indices of the heart in the patients. On the other hand, in the studies performed so far, a significant impact of antihypertensive drugs on RCP has been found, which explains approximately 29% of RCP variability (Lubas et al. 2018). Such a relation of renal hemodynamics may be reduced in patients who are not on hypotensive treatment and have preserved renal autoregulation. The active antihypertensive therapy should not be considered an undesirable factor, because a vast majority of patients with cardiorenal disorders are undergoing cardioprotective or nephro-protective treatment, which largely coincides with antihypertensive therapy.

The present work has several limitations. Patients with a stable kidney disease during the preceding 3 months, without renal biopsy, were qualified for the study; so the etiology of the kidney damage was unknown. Nonetheless, patients were not on active immunosuppressive therapy and showed no symptoms of hypervolemia, which minimized the probability of active pathology of renal hemodynamics. Another limitation of the study was a greater number of men than women and the lack of comparison with a group of healthy individuals. Despite the limitations, we believe we have shown that the left ventricular systolic and diastolic indices are independently associated with renal cortex perfusion in patients with hypertension and chronic kidney disease. Left ventricular global longitudinal strain seems better correlated with renal cortical perfusion compared to left ventricle ejection fraction.

Acknowledgments Funded by grant no 331 from the Military Institute of Medicine in Warsaw, Poland.

Conflicts of Interest The authors declare no conflict of interest in relations to this article.

Ethical Approval All procedures performed in studies involving human participants were in accordance with the ethical standards of the institutional and/or national research committee and with the 1964 Helsinki declaration and its later amendments or comparable ethical standards.

The study protocol was approved by the Bioethics Committee of the Military Institute of Medicine (No. 35/WIM/2011 of June 15, 2016).

Informed Consent Written informed consent was obtained from all individual participants included in the study.

References

Adamo L, Perry A, Novak E, Makan M, Lindman BR, Mann DL (2017) Abnormal global longitudinal strain predicts future deterioration of left ventricular function in heart failure patients with a recovered left ventricular ejection fraction. Circ Heart Fail 10(6):pii: e003788

Attias D, Macron L, Dreyfus J, Monin JL, Brochet E, Lepage L, Hekimian G, Iung B, Vahanian A, Messika-Zeitoun D (2013) Relationship between longitudinal strain and symptomatic status in aortic stenosis. J Am Soc Echocardiogr 26:868–874

Burke M, Pabbidi MR, Farley J, Roman RJ (2014) Molecular mechanisms of renal blood flow autoregulation. Curr Vasc Pharmacol 12:845–858

Choi JO, Cho SW, Song YB, Cho SJ, Song BG, Lee SC, Park SW (2009) Longitudinal 2D strain at rest predicts the presence of left main and three vessel coronary artery disease in patients without regional wall motion abnormality. Eur J Echocardiogr 10:695–701

Devereux RB, Alonso DR, Lutas EM, Gottlieb GJ, Campo E, Sachs I, Reichek N (1986) Echocardiographic assessment of left ventricular hypertrophy: comparison to necropsy findings. Am J Cardiol 57:450–458

Digne-Malcolm H, Frise MC, Dorrington KL (2016) How do antihypertensive drugs work? Insights from studies of the renal regulation of arterial blood pressure. Front Physiol 7:320

Gnanaraj JF, von Haehling S, Anker SD, Raj DS, Radhakrishnan J (2013) The relevance of congestion in the cardio-renal syndrome. Kidney Int 83:384–391

King A, Thambyrajah J, Leng E, Stewart MJ (2016) Global longitudinal strain: a useful everyday measurement? Echo Res Pract 3:85–93

Lang RM, Bierig M, Devereux RB, Flachskampf FA, Foster E, Pellikka PA, Picard MH, Roman MJ, Seward J, Shaneweise J, Solomon S, Spencer KT, St John Sutton M, Stewart W (2006) Recommendations for chamber quantification. Eur J Echocardiogr 7:79–108

Levey AS, Stevens LA, Schmid CH, Zhang YL, Castro AF, Feldman HI, Kusek JW, Eggers P, Van Lente F, Greene T, Coresh J, Disease-Epi CK (2009) A new equation to estimate glomerular filtration rate. Ann Intern Med 150:604–612

Liu YW, Tsai WC, Su CT, Lin CC, Chen JH (2009) Evidence of left ventricular systolic dysfunction detected by automated function imaging in patients with heart failure and preserved left ventricular ejection fraction. J Card Fail 15:782–789

Ljungman S, Laragh JH, Cody RJ (1990) Role of the kidney in congestive heart failure. Relationship of cardiac index to kidney function. Drugs 39(Suppl 4):10–21

Lubas A, Ryczek R, Kade G, Smoszna J, Niemczyk S (2013) Impact of cardiovascular organ damage on cortical renal perfusion in patients with chronic renal failure. Biomed Res Int 2013:137868

Lubas A, Ryczek R, Kade G, Niemczyk S (2015) Renal perfusion index reflects cardiac systolic function in chronic cardio-renal syndrome. Med Sci Monit 21:1089–1096

Lubas A, Kade G, Ryczek R, Banasiak P, Dyrla P, Szamotulska K, Schneditz D, Niemczyk S (2017) Ultrasonic evaluation of renal cortex arterial area enables differentiation between hypertensive and glomerulonephritis-related chronic kidney disease. Int Urol Nephrol 49:1627–1635

Lubas A, Kade G, Saracyn M, Niemczyk S, Dyrla P (2018) Dynamic tissue perfusion assessment reflects associations between antihypertensive treatment and renal cortical perfusion in patients with chronic kidney disease and hypertension. Int Urol Nephrol 50:509–516

Meyrier A (2015) Nephrosclerosis: update on a centenarian. Nephrol Dial Transplant 30:1833–1841

Mosteller RD (1987) Simplified calculation of body surface area. N Engl J Med 317:1098

National Kidney Foundation (2002) K/DOQI clinical practice guidelines for chronic kidney disease: evaluation, classification, and stratification. Am J Kidney Dis 39(Suppl 1):S1–S266

Ray P, Le Manach Y, Riou B, Houle TT (2010) Statistical evaluation of a biomarker. Anesthesiology 112:1023–1040

Romano S, Mansour IN, Kansal M, Gheith H, Dowdy Z, Dickens CA, Buto-Colletti C, Chae JM, Saleh HH, Stamos TD (2017) Left Ventricular global longitudinal strain predicts heart failure readmission in acute decompensated heart failure. Cardiovasc Ultrasound 15(1):6

Sahn DJ, DeMaria A, Kisslo J, Weyman A (1978) Recommendations regarding quantitation in M-mode echocardiography: results of a survey of echocardiographic measurements. Circulation 58:1072–1083

Scholbach T, Dimos I, Scholbach J (2004) A new method of color Doppler perfusion measurement via dynamic sonographic signal quantification in renal parenchyma. Nephron Physiol 96:99–104

Smiseth OA, Torp H, Opdahl A, Haugaa KH, Urheim S (2016) Myocardial strain imaging: how useful is it in clinical decision making? Eur Heart J 37:1196–1207

Yingchoncharoen T, Agarwal S, Popović ZB, Marwick TH (2013) Normal ranges of left ventricular strain: a meta-analysis. J Am Soc Echocardiogr 26:185–191

Adv Exp Med Biol - Clinical and Experimental Biomedicine (2019) 4: 9–18
https://doi.org/10.1007/5584_2018_283
© Springer Nature Switzerland AG 2018
Published online: 16 October 2018

Coupling of Blood Pressure and Subarachnoid Space Oscillations at Cardiac Frequency Evoked by Handgrip and Cold Tests: A Bispectral Analysis

Marcin Gruszecki, Yurii Tkachenko, Jacek Kot,
Marek Radkowski, Agnieszka Gruszecka, Krzysztof Basiński,
Monika Waskow, Wojciech Guminski, Jacek Sein Anand,
Jerzy Wtorek, Andrzej F. Frydrychowski, Urszula Demkow,
and Pawel J. Winklewski

Abstract

The aim of the study was to assess blood pressure–subarachnoid space (BP–SAS) width coupling properties using time–frequency bispectral analysis based on wavelet transforms during handgrip and cold tests. The experiments were performed on a group of 16 healthy subjects (F/M; 7/9) of the mean age 27.2 ± 6.8 years and body mass index of 23.8 ± 4.1 kg/m^2. The sequence of challenges

M. Gruszecki and A. Gruszecka
Department of Radiology Informatics and Statistics,
Faculty of Health Sciences, Medical University of Gdansk,
Gdansk, Poland

Y. Tkachenko and J. Kot
National Center for Hyperbaric Medicine, Faculty of
Health Sciences, Medical University of Gdansk, Gdynia,
Poland

M. Radkowski
Department of Immunopathology of Infectious and
Parasitic Diseases, Medical University of Warsaw,
Warsaw, Poland

K. Basiński
Department of Quality of Life Research, Faculty of Health
Sciences, Medical University of Gdansk, Gdansk, Poland

M. Waskow
Department of Clinical Anatomy and Physiology, Faculty
of Health Sciences, Pomeranian University of Slupsk,
Slupsk, Poland

W. Guminski
Department of Computer Communications, Faculty of
Electronics, Telecommunications and Informatics, Gdansk
University of Technology, Gdansk, Poland

J. S. Anand
Department of Clinical Toxicology, Faculty of Health
Sciences, Medical University of Gdansk, Gdansk, Poland

J. Wtorek
Department of Biomedical Engineering, Faculty of
Electronics, Telecommunications and Informatics, Gdansk
University of Technology, Gdansk, Poland

A. F. Frydrychowski
Department of Human Physiology, Faculty of Health
Sciences, Medical University of Gdansk, Gdansk, Poland

U. Demkow (✉)
Department of Laboratory Diagnostics and Clinical
Immunology of Developmental Age, Medical University
of Warsaw, Warsaw, Poland
e-mail: demkow@litewska.edu.pl

P. J. Winklewski
Department of Clinical Anatomy and Physiology, Faculty
of Health Sciences, Pomeranian University of Slupsk,
Slupsk, Poland

Department of Human Physiology, Faculty of Health
Sciences, Medical University of Gdansk, Gdansk, Poland

was first handgrip and then cold test. The handgrip challenge consisted of a 2-min strain, indicated by oral communication from the investigator, at 30% of maximum strength. The cold test consisted of 2 min of hand immersion to approximately wrist level in cold water of 4 °C, verified by a digital thermometer. Each test was preceded by 10 min at baseline and was followed by 10-min recovery recordings. BP and SAS were recorded simultaneously. Three 2-min stages of the procedure, baseline, test, and recovery, were analyzed. We found that BP–SAS coupling was present only at cardiac frequency, while at respiratory frequency both oscillators were uncoupled. Handgrip and cold test failed to affect BP–SAS cardiac–respiratory coupling. We showed similar handgrip and cold test cardiac bispectral coupling for individual subjects. Further studies are required to establish whether the observed intersubject variability concerning the BP–SAS coupling at cardiac frequency has any potential clinical predictive value.

Keywords

Bispectral analysis · Blood pressure · Cardiac frequency · Cold test · Handgrip test · Subarachnoid space width

1   Introduction

The handgrip test and the cold test are widely used to assess the stress response and activation of the sympathetic nervous system in humans. Nevertheless, the neural circuits involved in information processing differ between these tests. Particularly, neural processing of handgrip involves motor components, while cold test involves pain components (Vaegter et al. 2014; Macey et al. 2012). We and others also observed significant differences in cerebral blood volume and subarachnoid space (SAS) width responses to handgrip and cold tests (Winklewski et al. 2015a, b; Wilson et al. 2005). SAS width can be consid-

ered an indirect marker of cerebrospinal fluid volume (CSF) (Gruszecki et al. 2018b).

Both stress and pain are associated with white matter dysfunction and changes in neural plasticity and overall neural architecture (Coppieters et al. 2018; Bishop et al. 2017; Nugent et al. 2015; Sheikh et al. 2014; Upadhyay et al. 2012). One of the possible mechanisms that might be involved in white matter alterations is abnormal CSF pulsatility (Beggs et al. 2016a, b; Bateman et al. 2008). CSF pulsatile flow is driven by heart- and lung-generated blood inflows and outflows to and from the brain (Gruszecki et al. 2018a, b; Shi et al. 2018).

Periodic oscillations in CSF volume can be indirectly assessed using SAS width as a surrogate. In short, the main assumption for near-infrared transillumination/backscattering   sounding (NIR-T/BSS), the technique we use, is that translucent CSF in SAS acts as a propagation duct for infrared radiation, a technique resembling optical fiber engineering. This allows for measurement of SAS width to estimate changes in CSF volume (Frydrychowski and Pluciński 2007; Pluciński and Frydrychowski 2007; Pluciński et al. 2000). NIR-T/BSS has been validated against magnetic resonance imaging, showing comparable SAS width alterations induced by shifts in the body position (Frydrychowski et al. 2012).

The handgrip test results in increased blood pressure (BP) (Macey et al. 2012; Wszedybyl–Winklewska et al. 2012). Therefore, it seemed logical that relatively quick BP elevation may result in substantial alterations in the BP–SAS relationship. Quite surprisingly, analysis of amplitude coherences between these two signals has not revealed any substantial changes. On the contrary, sympathetic nervous system activation seems to stabilize the BP–SAS relationship (Winklewski et al. 2015a, b). Nevertheless, a lack of change in the amplitude coherence does not preclude alterations in the BP–SAS coupling detected with other mathematical methods.

The aim of the study was to assess the BP–SAS coupling properties using time–frequency bispectral analysis based on wavelet transforms (Clemson et al. 2016; Jamšek et al. 2004, 2007).

Consequently, our goal was to track time variability in coupling between these two oscillators at cardiac and respiratory frequencies. We hypothesized that both handgrip and cold tests will show substantial intersubject heterogeneity in the BP–SAS coupling. Nevertheless, as sympathetic nervous activation seems to dominate in terms of inter-signal behavior, we expected a similar handgrip and cold test bispectral coupling in individual volunteers.

2 Methods

2.1 Experimental Design

Experiments were performed on a group of 16 healthy volunteers, aged 27.2 ± 6.8 years and BMI of 23.8 ± 4.1 kg/m^2 (F/M; 7/9); none of them were smokers. All the subjects received detailed information about the study objectives and any potential adverse reactions. Although none of the participants suffered from known disorders or were taking any medication, a general and neurological examination was performed before the experiment. Nicotine, coffee, tea, cocoa, and methylxanthine-containing food and beverages were not permitted for 8 h before the tests. Additionally, prior to each test, the subjects were asked to rest comfortably for 30 min in the supine position.

All tests were conducted breathing ambient air at room temperature of 21 °C. The sequence of challenges was first handgrip and then cold test. For the handgrip challenge, subjects were instructed to squeeze an electronic dynamometer by the right hand at maximum force. They were initially directed to briefly squeeze at maximum effort as a reference. The challenge consisted of a 2-min strain, indicated by oral communication from the investigator, at 30% of the maximum. After the practice, subjects were allowed to return to a baseline state. The cold test consisted of 10 min at baseline, 2 min of hand immersion to the wrist level in the water of 4 °C, verified by a digital thermometer, and 10 min recovery. The investigator helped insert the hand into the water and take it out at the appropriate times.

2.2 Measurements

The mean BP was measured using continuous finger-pulse photoplethysmography (CNAP, CNSystems Medizintechnik AG, Graz, Austria). Finger blood pressure was calibrated against brachial arterial pressure. Oxyhemoglobin saturation (SaO$_2$) was measured continuously (Massimo Oximeter, Massimo, Milan, Italy) with a finger-clip sensor. Expired air was analyzed with the spirometry module of a medical monitoring system (Datex-Ohmeda, GE Healthcare, Wauwatosa, WI) for respiratory rate (RR) and minute ventilation (MV). Gas samples from the mouthpiece were constantly analyzed using the sidestream technique for end-tidal CO$_2$ (EtCO$_2$) and end-tidal O$_2$ (EtO$_2$) with the metabolic module of the same medical monitoring system. The NIR-T/BSS SAS signal was recorded with a head-mounted SAS 100 Monitor (NIRTI SA, Wierzbice, Poland). The theoretical and practical foundations of the NIR-T/BSS method have been published previously (Frydrychowski and Pluciński 2007; Frydrychowski et al. 2002). All variables were recorded continuously or videotaped, and the signals were digitally saved on the computer for further analysis.

2.3 Bispectral Analysis

Bispectrum analysis provides information about the coupling properties between interacting oscillators. The bispectrum is a frequency–frequency domain method that arises from higher-order statistics (Jamšek et al. 2004). However, the frequency–frequency domain is still unable to track time variability. Therefore, similarly to the time–frequency analysis, time–frequency–frequency analysis leads to a proposal of wavelet-based bispectral analysis (Jamšek et al. 2004, 2007) given by:

$$B(f_1, f_2) = \int W_T(f_1, t) W_T(f_2, t) W_T^*(f_3, t) dt$$

where $f_3 = 1/\left(\frac{1}{f_1} + \frac{1}{f_2}\right)$ is the wavelet coefficient and "*" denotes complex conjugation.

Additionally, to introduce the time-dependent variable, it is possible to define the bispectrum amplitude:

$$A(f_1, f_2, t) = | W_T(f_1, t) W_T(f_2, t) W_T^*(f_3, t) |$$

and bispectrum phase:

$$\phi(f_1, f_2, t) = \phi(f_1, t) + \phi(f_2, t) - \phi(f_3, t).$$

To be able to conclude that coupling between two oscillations at f_1 and f_2 exists, two conditions must be fulfilled (Clemson et al. 2016):

- A constant biphase $\phi(f_1, f_2, t)$ during at least ten periods of the lower-frequency interacting component.
- Peaks in the wavelet bispectrum $B(f_1, f_2)$ must be present at the same time as the biphase plateau.

In this study we were interested in finding a coupling between oscillations from different signals. The cross-bispectrum can be defined according to Clemson et al. (2016):

$$B_{122}(f_1, f_2) = \int W_1(f_1, t) W_2(f_2, t) W_2^*(f_3, t) dt$$

where W_1 and W_2 are the wavelet transforms of the corresponding time series.

It is also possible to calculate the cross-bispectrum from different combinations of signals:

$$B_{211}(f_1, f_2), B_{111}(f_1, f_2) \text{ or } B_{222}(f_1, f_2).$$

In this way we can obtain information about the direction of coupling between the oscillations in two different signals. Additionally, to remove the effect that the amplitude of the bispectrum is dependent on the amplitude of oscillations in the wavelet transform, a normalized cross-bispectrum can be defined as (Clemson et al. 2016):

$$b_{122}(f_1, f_2) = \frac{| B_{122}(f_1, f_2) |}{\sqrt{\int |W_2(f_1, t) W_2(f_2, t)|^2 dt \int |W_1(f_3, t)|^2 dt}}$$

2.4 Signal Processing

Three signals were recorded simultaneously from 16 subjects: blood pressure (BP), SAS_{LEFT} (SAS width in the left hemisphere), and SAS_{RIGHT} (SAS width in the right hemisphere). We analyzed three 2-min stages (total time 6 min) of the procedure: baseline, test, and recovery. The signals were resampled at 50 Hz. This gave an opportunity to analyze cardiac and respiratory frequency bands (Gruszecki et al. 2018a; Stefanovska et al. 1999). The characteristic cardiac (CF) and respiratory (RF) frequencies were identified: for each subject, the wavelet power spectrum was computed to identify those characteristic frequencies. Although the characteristic frequencies differ from subject to subject, they all lie within the defined frequency bands, defined by Stefanovska et al. (1999). Then, the wavelet cross-bispectrum was calculated as illustrated in Fig. 1. The value was calculated for the whole frequency domain and the entire duration of the procedure for one subject. It is apparent that the highest peak is located at the bifrequency (0.904 Hz, 0.904 Hz) belonging to the cardiac–cardiac (CF–CF) interaction. The lower peaks correspond to (0.22 Hz, 0.22 Hz) respiratory–respiratory (RF–RF) and (0.904 Hz, 0.22 Hz) cardiac–respiratory (CF–RF) interactions.

2.5 Statistical Analysis

The Wilcoxon signed-rank test was used to compare the changes in the median of the time-averaged measured values during the experiment (see Tables 1 and 2) for all the subjects. We compared the baseline to both tests' values.

Fig. 1 Normalized wavelet cross-bispectrum $\mathbf{b_{122}}$ analysis for the two simultaneously measured signals: blood pressure (BP) (1) and subarachnoid space (SAS$_{LEFT}$) (2) for one of the subject. Similar results we obtained for $\mathbf{b_{211}}$. The plot was prepared for the cold test for the entire duration of the procedure. Red (blue) dashed lines correspond to cardiac (respiratory) frequency

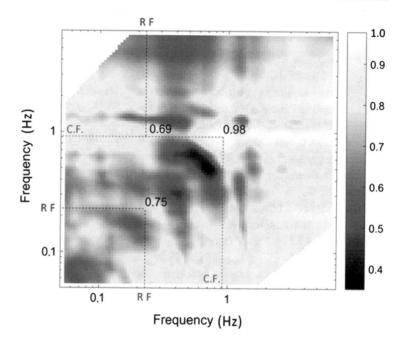

Table 1 Effects of 2-min cold test on the measured signals

Variable	Baseline	Cold test	p*
BP (mmHg)	120 ±16	121 ±15	ns
SAS$_{LEFT}$ (AU)	1.43 ±0.44	1.49 ±0.46	0.010
SAS$_{RIGHT}$ (AU)	1.62 ±1.16	1.66 ±1.17	ns
HR (beats/min)	64 ±7	72 ±8	0.003
SaO$_2$ (%)	99.1 ±1.4	99.4 ±1.2	ns
End-tidal CO$_2$ (kPa)	5.3 ±0.6	5.2 ±0.5	ns
End-tidal O$_2$ (kPa)	14.6 ±0.6	15.4 ±0.5	0.020
Respiratory rate (breaths/min)	15 ±5	16 ±4	0.040
MV (L/min)	7.3 ±1.3	8.9 ±1.5	0.006

Data are means ±SD. *BP* blood pressure, *SAS* subarachnoid space width, *AU* arbitrary units, *HR* heart rate, *SaO$_2$* oxyhemoglobin saturation, *CO$_2$* carbon dioxide, *O$_2$* oxygen, *MV* minute ventilation, p* for baseline *vs.* cold test difference, *ns* nonsignificant

Table 2 Effects of 2-min handgrip test on the measured signals

Variable	Baseline	Handgrip test	p*
BP (mmHg)	121 ±14	129 ±17	0.003
SAS$_{LEFT}$ (AU)	1.43 ±0.44	1.39 ±0.39	ns
SAS$_{RIGHT}$ (AU)	1.57 ±1.07	1.52 ±1.06	0.030
HR (beats/min)	64 ±7	75 ±19	0.001
SaO$_2$ (%)	97.4 ±2.1	97.9 ±1.9	ns
End-tidal CO$_2$ (kPa)	4.9 ±0.8	4.7 ±0.6	ns
End-tidal O$_2$ (kPa)	15.1 ±0.7	15.6 ±0.6	0.040
Respiratory rate (breaths/min)	14 ±4	17 ±4	0.007
MV (L/min)	8.4 ±1.9	9.1 ±1.2	0.003

Data are means ±SD. *BP* blood pressure, *SAS* subarachnoid space width, *AU* arbitrary units, *HR* heart rate, *SaO$_2$* oxyhemoglobin saturation, *CO$_2$* carbon dioxide, *O$_2$* oxygen, *MV* minute ventilation, p* for baseline *vs.* handgrip test difference, *ns* nonsignificant

3   Results

The time biphase was calculated at the defined bifrequency peaks (CF–RF, RF–RF, and CF–CF). Figure 2 shows the bispectral phase corresponding to the two peaks of the frequency pair CF–RF (a) and RF–RF (b) from Fig. 1. It is apparent that the biphases for both bifrequencies are not constant over time, which causes a lack of coupling for those pairs of bifrequencies. For all our subjects, we observed the same biphasic pattern for the pairs of bifrequencies.

We also estimated the time evolution of a biphase for the highest peak from Fig. 1, which corresponds to CF–CF bifrequencies. Figure 3 (panel a) illustrates the results obtained. The biphase for the whole time interval (baseline-

test-recovery) remained constant, and we observed a clear coupling for those pairs of bifrequencies. We observed a similar behavior for 10 out of the 16 subjects. For 2 out of the 16, the biphase was constant for the baseline and recovery stages but not for the test stage. This case is shown in Fig. 3b. In 4 out of the 16 subjects, the biphase for the whole procedure (no coupling) was variable (see Fig. 3c). Each subject showed the same behavior for both tests (handgrip and cold) and for the cross-bispectrum for the other combination of signals: BP and SAS_{RIGHT}. Descriptive statistics of recorded values are provided in Tables 1 and 2. The direction of change was very similar to those in our previous study (Winklewski et al. 2015b).

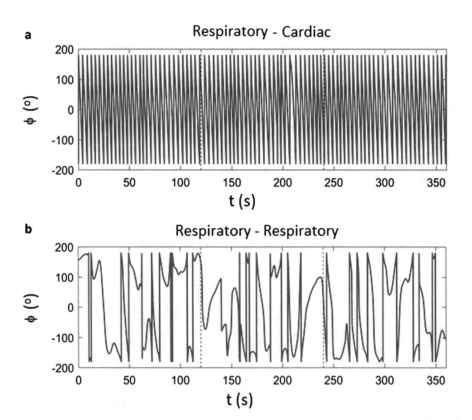

Fig 2 Bispectral phase corresponding to the frequency pair (**a**) cardiac–respiratory and (**b**) respiratory–respiratory. The values were estimated for the same subject as shown in Fig. 1. Red dashed lines show three stages of procedures: baseline, test, and recovery

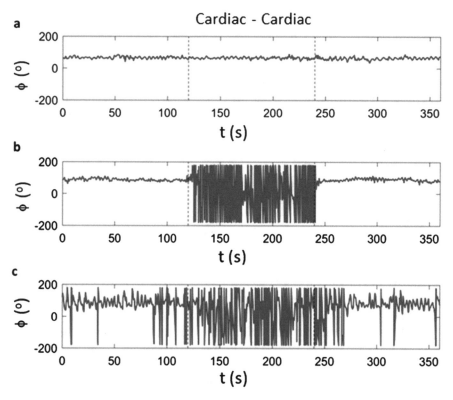

Fig. 3 Bispectral phase corresponding to the frequency pair cardiac–cardiac (**a–c**). The values in panel (**a**) were estimated for the same subject as Fig. 1. Red dashed lines show three stages of procedures: baseline, test, and recovery. For both handgrip and cold tests, and for each subject, a bispectral phase showed the same behavior

4　Discussion

There were two main findings of the study: (1) blood pressure–subarachnoid space (BP–SAS) width coupling was present only at cardiac frequency and (2) handgrip and cold tests exerted the same results concerning the BP–SAS coupling in individual subjects.

Heart-generated blood flow is pulsatile and needs to be dampened before reaching the microcirculation. In the aorta and large vessels, the Windkessel mechanism develops to store energy created by the systolic phase of the cardiac cycle. Consequently, from the thoracic aorta down to the end of arterioles, pulsatility is progressively attenuated (Safar and Struijker–Boudier 2010).

Brain microcirculation is additionally protected by CSF that absorbs abrupt systolic blood inflow, smooths capillary blood flow, and facilitates diastolic jugular outflow (Bateman et al. 2008). Therefore, CSF pulsatility is largely dependent on blood flow pulsatility. CSF pulsatile flow, in turn, is indispensable for the proper functioning of the glymphatic system, including utilization of metabolites and protein wastes of neuronal origin (Plog and Nedergaard 2018). Consequently, a better understanding of the relationship between BP and CSF (SAS) oscillations is of vital clinical and scientific significance.

Except for two subjects, handgrip and cold tests did not produce any changes in the BP–SAS coupling. Therefore, the results of the study are in line with our previous report that neither handgrip nor cold test influences the amplitude coherence between the analyzed signals (Winklewski et al. 2015a, b). However, the amplitude coherence does not provide any information about coupling but only about the relationship between the amplitude of modes in the signals measured. The bispectral analysis, in turn, provides information about the coupling between modes in the signals. Additionally, we can learn about the direction and strength of coupling. Therefore, bispectral analysis is a very robust mathematical tool, which provides us with useful information about signal properties.

The sympathetic nervous system may potentially interfere with the BP–SAS coupling in several ways. The role of sympathetic innervation in cerebral blood flow regulation remains a matter of controversy. Nevertheless, evidence accumulates that the sympathetic system may actually protect the brain from rapid BP surges (Wszedybyl–Winklewska et al. 2018; Winklewski and Frydrychowski 2013; Cassaglia et al. 2008, 2009). The link between cerebral homeostasis and BP seems, however, to be bidirectional. Most likely, even incremental alterations in intracranial pressure, for instance, those evoked by CSF pulsatile flow, may result in sympathetic activity changes and subsequent BP fluctuations (McBryde et al. 2017). Our present study supports the notion that the sympathetic system stabilizes the BP–SAS oscillation relationship.

Quite interestingly, in 25% of the subjects in this study (4 out of the 16), BP and SAS signals were not coupled with cardiac frequency. On the basis of the available data, we cannot explain this phenomenon. A larger study including patients with autonomic and central nervous system impairments might be needed to verify whether a lack of coupling between BP and CSF oscillations has any predictive clinical value. Importantly, both handgrip and cold tests only changed the coupling pattern at cardiac frequency in two subjects. Therefore, most likely, neither moderate static exercise nor cold affects the relationship between heart-generated BP and CSF signals. As we did not observe any differences between the handgrip and cold tests, we may speculate that sympathetic activation is the predominant factor determining the inter-signal behavior.

Breathing is another physiological process that affects both BP and SAS (Gruszecki et al. 2018a, b; Wszedybyl-Winklewska et al. 2017). Respiratory-driven BP and CSF fluctuations are largely of a mechanical nature. We have previously demonstrated clear coherence peaks in the respiratory interval, using a phase difference analysis, where BP and SAS signals are independent and oscillations are most likely generated centrally by the lungs (Gruszecki et al. 2018b). Our current study largely confirms those findings. Respiratory BP and SAS oscillations are not coupled to each other, and this lack of coupling is affected neither by handgrip nor by cold test.

A high within- and between-subject reproducibility and repeatability of NIR-T/BSS measurements have been demonstrated previously (Wszedybyl–Winklewska et al. 2011; Frydrychowski et al. 2002). Measurements with the use of infrared light (NIRS and NIR-T/BSS) do not allow for direct comparisons between subjects due to differences in skull bone parameters (Wagner et al. 2003). This limitation was, however, irrelevant in this study. NIR-T/BSS shows clear advantages for direct within-subject comparisons and signal analysis studies due to a high sampling frequency (Gruszecki et al. 2018a, b). As long as relative changes are analyzed, a high between-subject reproducibility is observed. We did not observe any sex-related differences. Nevertheless, it should be noticed that the study was neither aimed at nor powered to detect such differences.

In conclusion, we demonstrate that BP–SAS coupling was present only in the cardiac interval, whereas at respiratory frequency both oscillators were uncoupled. Neither did handgrip nor cold test affect BP–SAS cardiac and respiratory coupling. Cardiac bispectral coupling due to handgrip and cold tests was similar in individual subjects. Further studies are required to establish whether the observed intersubject variability concerning the BP–SAS coupling at cardiac frequency would be of any potentially predictive value in a clinical setting.

Funding The study was supported by the Medical University of Gdansk and the Medical University of Warsaw, Poland.

Conflicts of Interest Drs. Andrzej F. Frydrychowski, Wojciech Guminski, and Pawel J. Winklewski are stakeholders in NIRTI SA, the producer of NIR-T/BSS subarachnoid space oscillation monitors. None of the other authors have any fiduciary interests in the research reported herein.

Ethical Approval The experimental protocol of this study was approved by the Ethics Committee of the Medical University of Gdansk in Poland. All procedures performed in the study were in accordance with the ethical standards of the institutional research committee and with the 1964 Helsinki declaration and its later amendments.

Informed Consent Informed consent was obtained from all individual participants included in the study.

References

Bateman GA, Levi CR, Schofield P, Wang Y, Lovett EC (2008) The venous manifestations of pulse wave encephalopathy: Windkessel dysfunction in normal aging and senile dementia. Neuroradiology 50:491–497

Beggs CB, Magnano C, Belov P, Krawiecki J, Ramasamy DP, Hagemeier J, Zivadinov R (2016a) Internal jugular vein cross–sectional area and cerebrospinal fluid pulsatility in the aqueduct of Sylvius: a comparative study between healthy subjects and multiple sclerosis patients. PLoS ONE 11:e0153960

Beggs CB, Magnano C, Shepherd SJ, Belov P, Ramasamy DP, Hagemeier J, Zivadinov R (2016b) Dirty–appearing white matter in the brain is associated with altered cerebrospinal fluid pulsatility and hypertension in individuals without neurologic disease. J Neuroimaging 26:136–143

Bishop JH, Shpaner M, Kubicki A, Clements S, Watts R, Naylor MR (2017) Structural network differences in chronic muskuloskeletal pain: beyond fractional anisotropy. NeuroImage. https://doi.org/10.1016/j.neuroimage2017.12.021

Cassaglia PA, Griffiths RI, Walker AM (2008) Sympathetic nerve activity in the superior cervical ganglia increases in response to imposed increases in arterial pressure. Am J Physiol Regul Integr Comp Physiol 294:R1255–R1261

Cassaglia PA, Griffiths RI, Walker AM (2009) Cerebral sympathetic nerve activity has a major regulatory role in the cerebral circulation in REM sleep. J Appl Physiol 106:1050–1056

Clemson P, Lancaster G, Stefanovska A (2016) Reconstructing time–dependent dynamics. Proc IEEE 104:223–241

Coppieters I, De Pauw R, Caeyenberghs K, Lenoir D, DeBlaere K, Genbrugge E, Meeus M, Cagnie B (2018) Differences in white matter structure and cortical thickness between patients with traumatic and idiopathic chronic neck pain: Associations with cognition and pain modulation? Hum Brain Mapp 39:1721–1742

Frydrychowski AF, Pluciński J (2007) New aspects in assessment of changes in width of subarachnoid space with near–infrared transillumination–backscattering sounding, part 2: clinical verification in the patient. J Biomed Opt 12:044016

Frydrychowski AF, Gumiński W, Rojewski M, Kaczmarek J, Juzwa W (2002) Technical foundations for noninvasive assessment of changes in the width of the subarachnoid space with near–infrared transillumination–backscattering sounding (NIR–TBSS). IEEE Trans Biomed Eng 49:887–904

Frydrychowski AF, Szarmach A, Czaplewski B, Winklewski PJ (2012) Subarachnoid space: new tricks by an old dog. PLoS ONE 7:e37529

Gruszecki M, Nuckowska MK, Szarmach A, Radkowski M, Szalewska D, Waskow M, Szurowska E, Frydrychowski AF, Demkow U, Winklewski PJ (2018a) Oscillations of subarachnoid space width as a potential marker of cerebrospinal fluid pulsatility. Adv Exp Med Biol 1070:37–47

Gruszecki M, Lancaster G, Stefanovska A, Neary JP, Dech RT, Guminski W, Frydrychowski AF, Kot J, Winklewski PJ (2018b) Human subarachnoid space width oscillations in the resting state. Sci Rep 8:3057

Jamšek J, Stefanovska A, McClintock PVE (2004) Nonlinear cardio–respiratory interactions revealed by time–phase bispectral analysis. Phys Med Biol 49:4407

Jamšek J, Stefanovska A, McClintock PVE (2007) Wavelet bispectral analysis for the study of interactions among oscillators whose basic frequencies are significantly time variable. Phys Rev 76:046221

Macey PM, Wu P, Kumar R, Ogren JA, Richardson HL, Woo MA, Harper RM (2012) Differential responses of the insular cortex gyri to autonomic challenges. Auton Neurosci 168:72–81

McBryde FD, Malpas SC, Paton JF (2017) Intracranial mechanisms for preserving brain blood flow in health and disease. Acta Physiol 219:274e87

Nugent KL, Chiappelli J, Sampath H, Rowland LM, Thangavelu K, Davis B, Du X, Muellerklein F, Daughters S, Kochunov P, Hong LE (2015) Cortisol reactivity to stress and its association with white matter integrity in adults with schizophrenia. Psychosom Med 77:733–742

Plog BA, Nedergaard M (2018) The glymphatic system in central nervous system health and disease: past, present, and future. Annu Rev Pathol 13:379–394

Pluciński J, Frydrychowski AF (2007) New aspects in assessment of changes in width of subarachnoid space with near–infrared transillumination/

backscattering sounding, part 1: Monte Carlo numerical modeling. J Biomed Opt 12:044015

Pluciński J, Frydrychowski AF, Kaczmarek J, Juzwa W (2000) Theoretical foundations for noninvasive measurement of variations in the width of the subarachnoid space. J Biomed Opt 5:291–299

Safar ME, Struijker–Boudier HA (2010) Cross–talk between macro– and microcirculation. Acta Physiol 198:417e30

Sheikh HI, Joanisse MF, Mackrell SM, Kryski KR, Smith HJ, Singh SM, Hayden EP (2014) Links between white matter microstructure and cortisol reactivity to stress in early childhood: evidence for moderation by parenting. Neuroimage Clin 6:77–85

Shi Y, Thrippleton MJ, Marshall I, Wardlaw JM (2018) Intracranial pulsatility in patients with cerebral small vessel disease: a systematic review. Clin Sci 132:157–171

Stefanovska A, Bračič M, Kvernmo HD (1999) Wavelet analysis of oscillations in the peripheral blood circulation measured by laser Doppler technique. IEEE Trans Biomed Eng 46:1230–1239

Upadhyay J, Anderson J, Baumgartner R, Coimbra A, Schwarz AJ, Pendse G, Wallin D, Nutile L, Bishop J, George E, Elman I, Sunkaraneni S, Maier G, Iyengar S, Evelhoch JL, Bleakman D, Hargreaves R, Becerra L, Borsook D (2012) Modulation of CNS pain circuitry by intravenous and sublingual doses of buprenorphine. NeuroImage 59:3762–3773

Vaegter HB, Handberg G, Graven–Nielsen T (2014) Similarities between exercise–induced hypoalgesia and conditioned pain modulation in humans. Pain 155:158–167

Wagner BP, Gertsch S, Ammann RA, Pfenninger J (2003) Reproducibility of the blood flow index as noninvasive, bedside estimation of cerebral blood flow. Intensive Care Med 29:196–200

Wilson TD, Shoemaker JK, Kozak R, Lee TY, Gelb AW (2005) Reflex–mediated reduction in human cerebral blood volume. J Cereb Blood Flow Metab 25:136–143

Winklewski PJ, Frydrychowski AF (2013) Cerebral blood flow, sympathetic nerve activity and stroke risk in obstructive sleep apnoea. Is there a direct link? Blood Press 22:27–33

Winklewski PJ, Barak O, Madden D, Gruszecka A, Gruszecki M, Guminski W, Kot J, Frydrychowski AF, Drvis I, Dujic Z (2015a) Effect of maximal apnoea easy–going and struggle phases on subarachnoid width and pial artery pulsation in elite breath–hold divers. PLoS ONE 10:e0135429

Winklewski PJ, Tkachenko Y, Mazur K, Kot J, Gruszecki M, Guminski W, Czuszynski K, Wtorek J, Frydrychowski AF (2015b) Sympathetic activation does not affect the cardiac and respiratory contribution to the relationship between blood pressure and pial artery pulsation oscillations in healthy subjects. PLoS ONE 10:e0135751

Wszedybyl–Winklewska M, Frydrychowski AF, Michalska BM, Winklewski PJ (2011) Effects of the Valsalva maneuver on pial artery pulsation and subarachnoid width in healthy adults. Microvasc Res 82:369–373

Wszedybyl–Winklewska M, Frydrychowski AF, Winklewski PJ (2012) Assessing changes in pial artery resistance and subarachnoid space width using a non–invasive method in healthy humans during the handgrip test. Acta Neurobiol Exp 72:80–88

Wszedybyl–Winklewska M, Wolf J, Swierblewska E, Kunicka K, Mazur K, Gruszecki M, Winklewski PJ, Frydrychowski AF, Bieniaszewski L, Narkiewicz K (2017) Increased inspiratory resistance affects the dynamic relationship between blood pressure changes and subarachnoid space width oscillations. PLoS ONE e0179503:12

Wszedybyl–Winklewska M, Wolf J, Szarmach A, Winklewski PJ, Szurowska E, Narkiewicz K (2018) Central sympathetic nervous system reinforcement in obstructive sleep apnoea. Sleep Med Rev 39:143–154

Adv Exp Med Biol - Clinical and Experimental Biomedicine (2019) 4: 19–33
https://doi.org/10.1007/5584_2018_292
© Springer Nature Switzerland AG 2018
Published online: 10 November 2018

Influence of Heart Rate, Age, and Gender on Heart Rate Variability in Adolescents and Young Adults

Mario Estévez-Báez, Claudia Carricarte-Naranjo, Javier Denis Jas-García, Evelyn Rodríguez-Ríos, Calixto Machado, Julio Montes-Brown, Gerry Leisman, Adam Schiavi, Andrés Machado-García, Claudia Sánchez Luaces, and Eduardo Arrufat Pié

Abstract

Key autonomic functions are in continuous development during adolescence which can be assessed using the heart rate variability (HRV). However, the influence of different demographic and physiological factors on HRV indices has not been fully explored in adolescents. In this study we aimed to assess the effect of age, gender, and heart rate on HRV indices in two age groups of healthy adolescents (age ranges, 13–16 and 17–20 years) and two groups of healthy young adults (21–24 and 25–30 years). We addressed the issue using 5-min ECG recordings performed in the sitting position in 255 male and female participants. Time, frequency, and informational domains of HRV were calculated. Changes in HRV indices were assessed using a multiple linear regression model to adjust for the effects of heart rate, age, and gender. We found that heart rate produced more significant effects on HRV indices than age or gender. There was a progressive reduction in HRV with increasing age. Sympathetic influence increased with age and parasympathetic influence

M. Estévez-Báez and C. Machado
Institute of Neurology and Neurosurgery, Ministry of Health, Havana, Cuba

C. Carricarte-Naranjo, A. Machado-García, and C. S. Luaces
Faculty of Biology, Havana University, Havana, Cuba

J. D. Jas-García
Center for Sports Research, Havana, Cuba

E. Rodríguez-Ríos
Latin-American School of Medicine, Havana, Cuba

J. Montes-Brown
Department of Medicine & Health Science, University of Sonora, Hermosillo, Mexico

G. Leisman (✉)
Faculty of Social Welfare and Health Sciences, University of Haifa, Haifa, Israel

National Institute for Brain and Rehabilitation Sciences, Nazareth, Israel

University of the Medical Sciences, Faculty 'Manuel Fajardo', Havana, Cuba
e-mail: g.leisman@alumni.manchester.ac.uk

A. Schiavi
Anesthesiology and Critical Care Medicine, Neurosciences Critical Care Division, Johns Hopkins Hospital, Baltimore, MD, USA

E. A. Pié
Institute of Basic and Preclinical Sciences "Victoria de Girón", Havana, Cuba

progressively decreased with age. The influence of gender was manifest only in younger adolescents and young adults. In conclusion, age, gender, and particularly heart rate have a substantial influence on HRV indices, which ought to be considered to avoid biases in the study of the autonomic nervous system development. The lack of the gender-related effects on HRV indices in late adolescence could be related to non-completely achieved maturity of the autonomic mechanisms, which deserves further exploration.

Keywords

Adolescents · Age · Autonomic nervous system · Gender · Heart rate variability · Power spectral analysis

1   Introduction

Sequential fluctuations of heart inter-beat intervals are the most ostensible evidence of the chronotropic cardiovascular regulation exerted by the autonomic nervous system. Heart rate variability (HRV) has become one of the most sensitive, noninvasive, and reliable assessment of the integrity and functional status of the autonomic nervous system (Mestanikova et al. 2016; Tonhajzerova et al. 2016; Sassi et al. 2015; Task Force of ESC and NASPE 1996; Kuusela 2013; Sosnowski 2011). Among many factors related to HRV, age and gender are the most recognized and studied (Almeida-Santos et al. 2016; Pothineni et al. 2016; Sharma et al. 2015; Abhishekh et al. 2013; Michels et al. 2013; Moodithaya and Avadhany 2012; Antelmi et al. 2004; Migliaro et al. 2001; Tsuji et al. 1994). Heart rate is considered a modifying factor of HRV, but its effect is not always considered in HRV studies (Voss et al. 2013, 2015; Kuo et al. 1999). Recently, different reports have emphasized the importance of heart rate in the HRV analysis (van Roon et al. 2016; Estévez-Báez et al. 2015a, b; Goldberger et al. 2014; Monfredi et al. 2014; Sacha 2013; Nieminen et al. 2007).

The physiological mechanisms during adolescence are actively and progressively changing. HRV can be used to ascertain the evolution of the ontogenetic maturation (Evans et al. 2016; Moodithaya and Avadhany 2012; Dogru et al. 2010; Fontani et al. 2004). Although HRV has been extensively used to explore the function of the autonomic nervous system in many different age ranges, the early (13–16 years) and late teenagers (17–20 years) have not been fully explored. Therefore, this study seeks to define the effects of the main factors known to influence HRV, such as heart rate, age, and gender in healthy adolescents and young adults.

2   Methods

2.1   Participants

A cohort of 255 healthy subjects of both genders was studied. There were two age groups of healthy adolescents of 13–16 and 17–20 years of age and another two age groups of healthy young adults of 21–24 and 25–30 years of age. To be included they had to willingly agree to participate in the study and to show a normal 12-lead electrocardiogram (ECG). Exclusion criteria were history of smoking, cardiorespiratory or neurological disorder, diabetes mellitus, and the use of medications with known autonomic nervous system effects. The subjects were categorized into four age groups: group A from 13 to 16, group B from 17 to 20, group C from 21 to 24, and group D from 25 to 30 years of age. The adolescents included were recruited from high schools and other educational centers near the National Institute of Neurology and Neurosurgery (INN) in Havana, Cuba. The young adults were recruited from physicians, medical students, nurses, and technicians from the INN and other nearby medical institutions.

2.2   Experimental Sessions

All subjects were studied from 08:00 a.m. to 12:00 p.m. They were instructed to abstain from physical efforts the day before the study, avoid caffeine, sleep for at least 7 h the night before, have their usual breakfast, and to drink a glass of fruit juice at least 1 h before the study. Body mass (accuracy 0.1 kg) and body height (accuracy 0.1 cm) were measured with standard clinical anthropometric instruments. Body mass index (BMI) was calculated as body mass (kg) divided by body height (m) squared. Female participants were studied in the mid-follicular phase of their menstrual cycle. All participants had to rest for 30 min sitting in a chair in a semi-reclining position while the ECG electrodes were placed, the equipment was set and calibrated, and during the check of recording quality. The temperature in the laboratory was maintained about 25 °C.

2.3   Electrocardiogram (ECG) Recordings

ECG were recorded for 15 min with commercial amplifiers (monitor Hewlett Packard 78354A, Palo Alto, CA) and digitized with a 12-bit analog-digital (A/D) converter board (USB-6008 DAQ, National Instruments, Austin, TX). A/D conversion was carried out with a sampling rate of 1 kHz. To control the process of digitization and storing of the ECG signal in the hard PC disk, specific software was developed by one of the authors (JDJG) and written in LabView v10.0 (National Instruments, Austin, TX). Filters were set for a band spectrum of 0.5–45 Hz. The ECG signal was obtained from disposable electrodes placed on the chest in positions CM2 and V5.

2.4   Electrocardiogram Analysis, Edition, and Preprocessing

ECG recordings were imported offline to a software tool developed in Delphi Embarcadero XP by the authors MEB and AMG (MultiTools v3.1.2, 2009–2016) for visual inspection and detection of the fiduciary "R" peaks. Accurate "R" peak automatic detections were visually checked and properly corrected, as required. Persons with isolated ventricular ectopic beats or supraventricular events were not included in this study. Five-minute segments of artefact free R-R

inter-beat intervals (RRi) were subject to the preprocessing procedures that included (a) RRi series resampling using an interpolation method applying cubic splines with a sampling frequency of 6.82 Hz, to transform the R-R sequences to evenly sampled time series; (b) R-R series demeaning, consisting of subtraction of the mean RR value from all RR items to significantly reduce the DC component of RRi series; (c) linear detrending, computing a least squares fit of a straight line sequence to the data and subtracting the resulting function from the RRi series; and (d) zero-phase-shift digital filtering of the RRi series using the order 6 Butterworth high-pass filter with a cutoff frequency of 0.02 Hz, to eliminate undesirable frequency components, without affecting the phase components.

2.5   HRV Indices

The following HRV indices were calculated in the time domain: the mean R-R inter-beat period duration (MRRi), the standard deviation of the normal RRi (SDNN), the root mean square of successive differences (RMSSD), and the variation coefficient ((SDNN/MRRi)*100) expressed as percent (CVar).

For the frequency domain indices, a total of 2048 samples (5-min RRi series) were used in the computation of the Welch modified periodogram with a Hamming window, using segments of 512 samples and overlapping periods of 256 samples (50%). A more detailed description of this method can be found elsewhere (Estévez-Báez et al. 2015a, b; Machado et al. 2014; Machado-Ferrer et al. 2013). The limits for the spectral HRV very low-frequency (VLF) band were 0.02–0.04 Hz. The low-frequency band (LF) was considered 0.04–0.15 Hz and the high-frequency (HF) band was 0.15–0.40 Hz. The absolute power spectral density estimations were calculated as the integral of each one-sided quadratic spectrogram in the frequency ranges previously defined and marked with the acronyms P_VLF, P_LF, P_HF, and P_Tot. The ratio of power spectra of LF-to-HF bands (LF/HF) also was calculated. The relative power content for the three specified bands was normalized as the corresponding percent of the total power spectral density of RRi series (P_Tot) and the following acronyms were used for the indices calculated: nu_VLF, nu_LF, and nu_HF. The complexity of the RRi series was calculated with the Shannon entropy index, considered the informational domain index (Bravi et al. 2011), using the expression:

$$H = - \sum_{i=1}^{N} p_i \log_2 p_i \qquad (1)$$

where p_i is the probability of every possible value of the RRi duration and N is the total number of samples.

Digital signal processing in this study was carried out using the custom-tailored programs developed by our staff with MATLAB (MathWorks v9.1.0.441655 R2016b).

2.6 Adjustment of Measured HRV Indices

The influence of age on HRV indices between the groups was investigated as the effect of the covariation of heart rate and gender, with the intent to adjust for those correlations if present. Likewise, the influence of gender was investigated as the effect of the covariation of heart rate and age. The influence of the heart rate on HRV indices was investigated as the effect of the covariation of the reciprocal values of heart rate, i.e., the mean heart inter-beat period duration (MRRi). Gender was coded as a binary variable with zero representing female and one representing male. A linear model was computed for the different HRV indices and was used for adjustments of heart rate and gender and for adjustments of heart rate and age in the following form:

$$y = \beta_0 + \beta_1 x_1 + \beta_2 x_2 + \beta_3 x_1 * x_2 \qquad (2)$$

where y is the adjusted HRV index; β_0, β_1, β_2, and β_3 are the regression coefficients of the factors included; and their interactions are represented by the product $x_1 * x_2$.

The linear model used for the evaluation of the effect of heart rate, represented by its reciprocal value, i.e., MRRi, was in the form:

$$y = \beta_0 + \beta_1 (x_1) \qquad (3)$$

The validity of the linear models was assessed using an F-test to prove the null hypothesis that the regression coefficients β_1, β_2, and β_3 were all equal zero or to say that the model was constant in the form:

$$y = \beta_0 \qquad (4)$$

Then, a t-test for each individual regression coefficient determined if the covariation of the factor associated with the HRV index was significant. If the F-statistics yielded a nonsignificant result at $p > 0.05$, then the HRV index did not require adjustment. Otherwise, the residuals of the linear model were calculated as the differences between the observed values and the values predicted by the model. If the adjustment was considered meaningful, its distribution was evaluated with the Kolmogorov-Smirnov test. For normal distribution, the calculated residuals became the adjusted values for the HRV index and were submitted to statistical comparison between the groups.

2.7 Statistical Analysis

The results are presented as means ±SD. For comparisons between age of male and female subjects in each of the four groups, a t-test was used. Possible differences in BMI between groups were tested using one-way ANOVA. Normality of data distribution was evaluated with the Shapiro-Wilk test. The non-normally distributed data were modified using natural or common logarithmic transformations. The parametric correlation measures of the Pearson product moment were used to test the strength of the relationships between heart rate and age with the HRV indices, before and after adjustments. The Spearman rank nonparametric correlation measures were used to test the relationships between gender and HRV indices.

A one way ANOVA was used for comparing HRV indices between the four age groups for the dimensions of: without adjustments, and after adjusting for MRRi and gender. The statistical power for each test was only considered valid for values over 0.7, and the *post-hoc* comparisons were conducted using the Scheffe test. To evaluate differences related to gender, after adjusting for heart rate and age, a factorial ANOVA was used followed by the *post-hoc* Duncan test. Significance was set at p < 0.05. All statistical analyses were performed using a commercial statistical package of Statistica v10 (StatSoft Inc., Tulsa, OK).

3 Results

The age and gender characteristics of the participants included in this study are presented in Table 1. There were no significant differences (F (3,251) = 1.968; $p = 0.12$) between the BMI values in the four age groups (Group A, 21.1 ± 2.2 kg/m^2; Group B, 21.1 ± 2.7 kg/m^2; Group C, 21.7 ± 2.9 kg/m^2; and Group D, 21.6 ± 2.8 kg/m^2). A nonlinear relationship was shown between the values of the mean heart period against the corresponding mean heart rate values (reciprocal values) for the 255 participants of this study (Fig. 1).

Table 1 Age and gender characteristics of the participants

Groups	Age range (years)	Males (n)	Females (n)	Whole group (n)	p
A	13–16	(30) 14.4 ± 0.7	(31) 14.6 ± 0.8	(61) 14.4 ± 0.8	0.237
B	17–20	(27) 18.2 ± 0.9	(38) 18.4 ± 0.9	(65) 18.4 ± 0.9	0.318
C	21–24	(31) 22.8 ± 1.2	(37) 22.6 ± 1.1	(68) 22.6 ± 1.2	0.413
D	25–30	(28) 28.3 ± 1.6	(33) 27.7 ± 1.5	(61) 28.0 ± 1.6	0.118
Total	13–30	(116) 21.3 ± 5.2	(139) 21.1 ± 4.7	(255) 21.2 ± 5.0	0.809

Data are means ±SD; *n* number of subjects; *p-value,* *t*-test for independent samples for age comparison between males and females

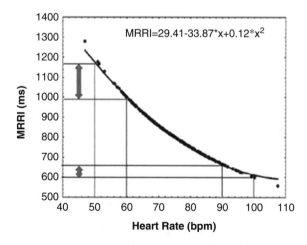

Fig. 1 Scatterplot diagram obtained by polynomial fitting of the values observed in the R-R tachograms of heart frequency and its corresponding heart period reciprocal values, in the group of 255 healthy adolescents and young adults. Points represent observed values and the continuous line depicts the fitting curve. Note the evident differences in RR interval ranges (vertical arrows) corresponding to increments of identical range of the heart rate values at lower (50–60 bpm) and higher values (90–100 bpm). MRRi, mean heart inter-beat period duration; bpm, beats per min

3.1 Comparisons of Original Heart Rate Variability (HRV) Indices Without Corrections

The strength of the relationship between different HRV indices and MRRi, age, and gender is substantiated by significant correlations of MRRi with 10 out of the 11 HVR indices (90.9%). There were significant correlations of the age factor with 6 HVR indices (54.5%), while the gender factor correlated with 4 HVR (36.4%) (Table 2).

The MRRi values showed significant differences between the younger adolescents (Group A) and both groups of adults (C and D) and between the older adolescents (Group B) and older adults (Group D) as shown in Table 3. The complexity Shannon's entropy index showed significant differences only between the adolescents of group A and the values of the other groups. The original values of the power spectral density, expressed as the absolute terms, showed differences only for the index P_HF between the younger adolescents (Group A) and the older adults (Group D) (Table 3). The indices of time domain variability SDNN and RMSSD did not show significant differences between age groups. The values of power spectral density in normalized units showed significant differences between Group A and Group C for the nu_LF index and for the LF/HF ratio.

3.2 Adjustment Procedures

Calculations carried out for the adjustment of some of the spectral HRV indices detailed in Methods (Sect. 2.6) are exemplified in Table 4. Data concern the adjustment for MRRi, age, and the interaction of the two. The F-tests show highly significant values for the four HRV indices and also indicate that the calculated residuals were distributed normally, checked with the Kolmogorov-Smirnov tests. According to the rules of the procedure, these residuals could then be considered as values statistically free of the effect of MRRi, age, and their interaction. In addition to this information, we can observe that not all the factors, or their interaction, significantly contributed to the results of the whole adjustment procedure. All the t-statistic values were significant for the beta estimates of the linear regression model for the P_LF index, but not entirely so for the other indices. This information may be used to better understand the strength of the interrelationships of heart rate, age, and gender included in the regression model.

3.3 Values Adjusted for the Mean R-R Inter-Beat Period Duration (MRRi)

The adjustment procedure shows that the nu_VLF index did not need to be adjusted, because the F-statistic was nonsignificant. After adjusting for the effect of MRRi, all HRV indices showed significant correlations with age, but the correlations with gender remained unchanged (Table 2).

There were significant differences between the age groups for all the absolute HRV power spectral indices except for the values of the VLF band. The time domain variability indices (SDNN, RMSSD) showed significant differences between age groups that were not detected before correction. The Shannon entropy complexity index showed significant differences between the values of the young adolescents (Group A) and the other three groups and between the older adolescents (Group B) and the young adults Groups C and D (Fig. 2). Significant differences between the adolescent Groups A and B were found for the indices P_HF, SDNN, and RMSSD. The correction for the LF/HF ratio showed significant differences between the younger adolescent Group A and younger adult Group C and also between Group A and older adult Group D.

Table 2 Correlations between HRV indices and the original values and after successive adjustments for MRRi, gender, and age

HRV index	Original values			Adjustment for MRRi		Adjustment for MRRi and gender	Adjustment for MRRi and age
	MRRi	Age	Gender	Age	Gender	Age	Gender
ln P_VLF (ms^2)	**0.38** (0.00..)	−0.05 (0.45)	0.10 (0.12)	**−0.17** (0.01)	0.05 (0.42)	**−0.18** (0.00..)	0.03 (0.61)
ln P_LF (ms^2)	**0.37** (0.00..)	−0.04 (0.58)	**0.18** (0.01)	**−0.15** (0.01)	**0.14** (0.03)	**−0.20** (0.00..)	0.11 (0.07)
ln P_HF (ms^2)	**0.41** (0.00..)	**−0.20** (0.01)	0.01 (0.89)	**−0.35** (0.00..)	−0.05 (0.44)	**−0.36** (0.00..)	−0.07 (0.26)
ln P_Tot (ms^2)	**0.46** (0.00)	−0.11 (0.09)	0.11 (0.07)	**−0.27** (0.00..)	0.06 (0.32)	**−0.27** (0.00..)	0.04 (0.53)
nu_VLF (%)	−0.05 (0.44)	0.11 (0.07)	0.02 (0.80)	0.13 (0.04)	0.02 (0.72)	0.13 (0.04)	0.03 (0.67)
nu_LF (%)	**−0.15** (0.02)	**0.21** (0.00..)	**0.16** (0.01)	**0.26** (0.00..)	**0.18** (0.01)	**0.27** (0.00..)	**0.18** (0.01)
nu_HF (%)	**0.15** (0.02)	**−0.23** (0.00..)	**−0.15** (0.02)	**−0.27** (0.00)	**−0.17** (0.01)	**−0.29** (0.00..)	**−0.17** (0.01)
ln LF/HF ratio (nu)	**−0.15** (0.02)	**0.22** (0.00..)	**0.16** (0.01)	**0.27** (0.00..)	**0.19** (0.01)	**0.28** (0.00..)	**0.19** (0.01)
log$_{10}$ SDNN (ms)	**0.51** (0.00..)	−0.11 (0.07)	0.11 (0.07)	**−0.31** (0.00..)	0.06 (0.37)	**−0.31** (0.00..)	0.04 (0.53)
log$_{10}$ RMSSD (ms)	**0.58** (0.00..)	**−0.18** (0.01)	0.02 (0.73)	**−0.43** (0.00..)	−0.07 (0.29)	**−0.45** (0.00..)	−0.09 (0.14)
Shannon entropy (cu)	**0.54** (0.00..)	**0.41** (0.00..)	0.07 (0.30)	**0.30** (0.00..)	−0.00 (0.94)	**0.29** (0.00..)	−0.03 (0.64)

Values are presented as Pearson's or Spearman's correlation coefficients and its associated probabilities shown in parenthesis. Abbreviations used for HRV indices are those described in Methods. Highlighted values are significant at $p < 0.05$; nu non-dimensional units, cu conventional units, ln natural logarithm; log$_{10}$ common logarithm (base 10); 0.00.. highly significant values for at least $p < 0.001$

Table 3 Statistical differences observed for the original heart rate variability (HRV) indices calculated in the four age groups of healthy adolescents and young adults (one-way ANOVA)

HRV index	Group A (n = 61)	Group B (n = 65)	Group C (n = 68)	Group D (n = 61)	F (3251)	p	Observed power (α = 0.05)
MRRi (ms)	781.43 ±92.70	814.75 ±107.80‡	838.46 ±107.20*	873.47 ±116.40*	8.16	0.00003	0.99
SDNN (ms)	65.40 ±22.70	61.50 ±23.30	60.51 ±20.30	58.55 ±23.80	–	–	–
Log₁₀ SDNN (ms)	1.79 ±0.15	1.76 ±0.15	1.76 ±0.15	1.74 ±0.17	1.19	0.31	0.31
RMSSD (ms)	61.43 ±32.70	53.53 ±29.20	47.63 ±24.01	50.33 ±36.10	–	–	–
Log₁₀ RMSSD (ms)	1.73 ±0.22	1.67 ±0.22	1.64 ±0.19	1.64 ±0.22	3.01	0.03	0.68
Shannon entropy (cu)	6.77 ±0.48	7.64 ±0.45**	7.61 ±0.45**	7.54 ±0.50**	48.28	0.0000	1.00
Ln P_VLF (ms²)	10.82 ±0.72	10.79 ±0.71	10.70 ±0.75	10.77 ±0.75	0.35	0.79	0.12
Ln P_LF (ms²)	12.65 ±0.78	12.61 ±0.70	12.67 ±0.75	12.53 ±0.77	0.44	0.73	0.14
Ln P_HF (ms²)	12.60 ±0.85	12.36 ±1.00	12.20 ±0.96	12.12 ±0.93*	3.23	0.02	0.74
Ln P_Tot (ms²)	13.45 ±0.74	13.34 ±0.75	13.31 ±0.73	13.22 ±0.74	1.01	0.39	0.27
nu_VLF (%)	8.12 ±4.10	8.89 ±4.30	8.71 ±4.20	9.93 ±4.40	1.51	0.21	0.40
nu_LF (%)	46.67 ±12.80	50.49 ±14.60	54.73 ±13.70**	53.17 ±13.30	3.79	0.01	0.81
nu_HF (%)	45.20 ±14.60	40.62 ±15.90	36.55 ±15.70*	36.90 ±17.10**	4.05	0.008	0.84
LF/HF ratio (nu)	1.26 ±0.76	1.63 ±1.09	2.06 ±1.53	2.05 ±1.71	–	–	–
Ln LF/HF ratio (nu)	0.04 ±0.63	0.25 ±0.72	0.47 ±0.73**	0.41 ±0.85	4.13	0.007	0.85

Values are means ±SD; transformed values are presented when it was necessary to achieve normality distributions. Abbreviations used for HRV indices are those described in Methods; *ln* natural logarithm; log₁₀ common logarithm (base 10); *cu,* conventional units, *cu* normalized units, *nu* non-dimensional units, *F* values of Fisher's statistics for (n, m) degrees of freedom, *p* associated probability for observed F values, *Observed Power* power analysis of the F statistics for an α error of 0.05 (valid results only were accepted for values over 0.7). Results for significant post hoc Scheffe's tests are indicated with symbols in the corresponding columns: *$p < 0.05$, **$p < 0.01$ Groups B, C, and D vs. Group A, ‡$p < 0.05$ Group B vs. Group D

3.4 Values Adjusted for the Effect of the Mean R-R Inter-Beat Period Duration (MRRi) and Gender

The nu_VLF index did not require adjustment, according to the F-statistic value. The adjustment produced significant values only for the correlation indices between the HRV indices and the

factor age (Table 2). The HRV indices that significantly differed between the age groups are shown in Fig. 3. The general feature of spectral indices, expressed as the absolute power, was a reduction of their values with increasing age. The LF/HF ratio increased with age. Low frequency, expressed in normalized units (nu_LF), showed an increment with age, while the opposite was observed for high frequency (nu_HF). Corrected

Table 4 Summary of calculations of the adjustment procedure applied to some indices of heart rate variability for MRRi, age, and their interaction in the investigated adolescents and young adult subjects ($n = 255$)

HRV index		Estimated coefficients of the linear regression model				F(3251)	p	K
		β estimates	SE	t-statistic	p			
P_VLF	Intercept	**6.527**	**1.416**	**4.61**	0.00	**18.92**	0.00..	0.055
	Age	0.085	0.064	1.33	0.184			p > 0.20
	MRRi	**0.006**	**0.002**	**3.36**	0.001			
	Age*MRRi	−0.0001	0.0001	−1.74	0.082			
P_LF	Intercept	**7.081**	**1.445**	**4.90**	0.00	**18.66**	0.00..	0.042
	Age	**0.147**	**0.065**	**2.26**	0.024			
	MRRi	**0.007**	**0.002**	**4.16**	0.00			p > 0.20
	Age*MRRi	**−0.0002**	**0.0001**	**−2.64**	0.009			
P_HF	Intercept	**8.307**	**1.722**	**4.82**	0.00	**32.99**	0.00	0.041
	Age	0.016	0.077	0.21	0.837			p > 0.20
	MRRi	**0.007**	**0.002**	**3.13**	0.002			
	Age*MRRi	−0.001	0.00001	−1.08	0.28			
P_Tot	Intercept	**8.431**	**1.344**	**6.28**	0.00	**33.17**	0.00	0.038
	Age	0.088	0.060	1.46	0.146			p > 0.20
	MRRi	**0.007**	**0.002**	**4.24**	0.00			
	Age*MRRi	**−0.0002**	**0.0001**	**−2.12**	0.034			

Abbreviations of HRV indices are those described in Methods. *F* values of Fisher's statistics for (n, m) degrees of freedom, *p* associated probability to the corresponding F-Statistic values, *K* value of the Kolmogorov-Smirnov test and associated probability for rejecting the hypothesis of normality of the calculated residuals; values in bold font correspond to significant results; *0.00*, highly significant values for at least $p < 0.00001$

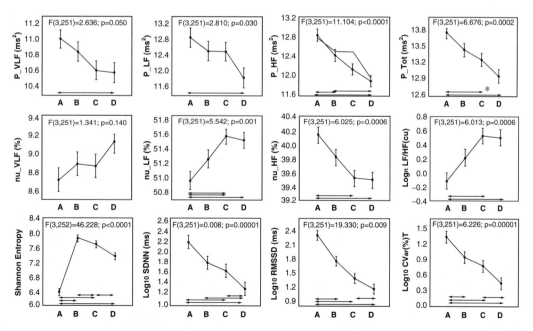

Fig. 2 Comparison of heart rate variability (HRV) indices in the two age groups of adolescents (A and B) and two age groups of adults (C and D) after adjusting for the mean R-R inter-beat period duration (MRRi). The results obtained from one-way ANOVA are specified in the upper part of each diagram. Vertical bars denote ±SE. Significant results of post hoc Scheffe tests are represented by double-arrow lines for $p < 0.01$. The symbol (*) indicates $p < 0.05$

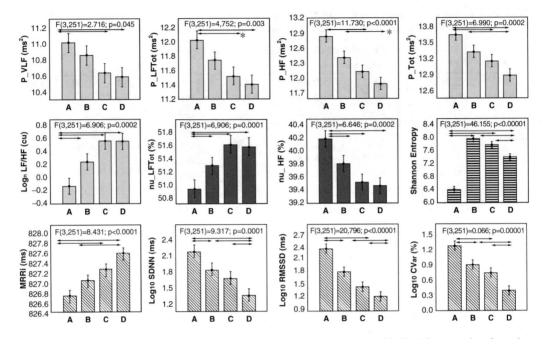

Fig. 3 Comparison of HRV indices showing significant differences between the two age groups of adolescents (A and B) and the two age groups of adults (C and D) after adjusting for the mean heart period and gender.

Vertical bars denote ±SE. Significant results of post hoc Scheffe tests are represented by double-arrow lines for $p < 0.01$. The symbol (*) indicates $p < 0.05$

MRRi values showed a progressive increment with age, while the variability indices SDNN and RMSSD evidently decreased with increasing age. Values of the Shannon entropy index showed significant differences between all the groups, with a sharp increase between the younger and older adolescents (Group A vs. Group B) and progressive significant reductions between Group B and both younger and older adults (Groups C and D).

were detected in both younger and older adult Groups C and D, with higher values for females. The LF/HF ratio was significantly higher for the male, compared to female, participants of Group D. The corrected MRRi index for age was significantly lower in female, compared to male, participants, only in the adolescents of Group A.

3.5 Values Adjusted for the Effect of MRRi and Age Interaction

The nu_VLF index did not require adjustment. Gender influence, after adjusting the original RRi series by MRRi and age, was found only in 4 out of the 11 HRV indices investigated (Fig. 4). For the nu_LF index, significant inter-gender differences were detected only in the older adult Group D, with higher values for males. For the nu_HF index, significant inter-gender differences

4 Discussion

The heart rate factor produced more significant effects on HRV than age and gender, modifying the original values in that the detection of differences between the participants of the four age groups was strongly reduced. There was a progressive reduction of HRV with increasing age, shown by the absolute indices of spectral power density in all the HRV bands and also for the total power. Sympathetic influence increased with age, which consisted of increases in the LF/HF ratio and in spectral power density expressed in normalized units for the

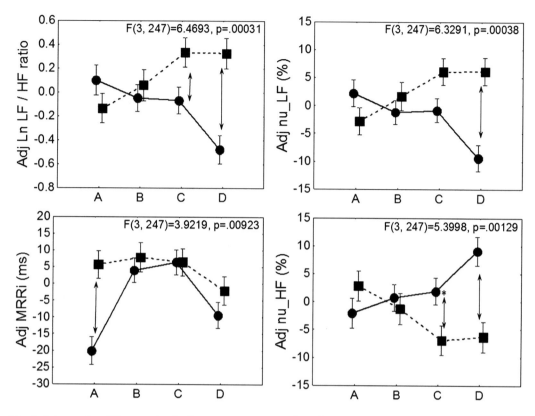

Fig. 4 Inter-gender differences in the four age groups after adjusting the HRV original values by age and mean heart period. Solid lines represent females and dashed lines represent males. Vertical bars denote ±SE. Significant results of post hoc Duncan tests are represented by double-arrow lines for $p < 0.01$. The symbol (*) indicates $p < 0.05$

low-frequency spectral band (P_LF), and a reduction in the global variability of the time domain index SDNN. Parasympathetic influence, on the other hand, progressively declined with age, which consisted of decreases in P_HF, nu_HF, and RMSSD. Nevertheless, the MRRi, an integral index of the central control of the autonomic nervous system on cardiac chronotropic activity, increased progressively with age, leading to a reduction in HR. The effect of gender on HRV consisted of significant differences in young adults (Groups C and D), demonstrating increments in the sympathetic HRV indices P_LF and nu_LF and a concomitant reduction in the parasympathetic nu_HF in males. The MRRi index was significantly lower, pointing to a higher heart rate in females, but only in the younger adolescents (Group A).

To the best of our knowledge, this is the first study to explore the autonomic cardiovascular control in adolescents between the ages of 13–16 and 17–20 years, compared with young adults of 21–24 and 25–30 years of age. Here, we assessed the influence of age, gender, and heart rate on HRV indices in these age groups. The relevance of the two age ranges of adolescents stems from the fact that in many cultures individuals aged 17–20 are chronologically and legally considered as mature enough to assume a number of responsibilities such as driving a vehicle, serving in the armed forces, voting, or marrying (Arnett 2007; Christie and Viner 2005; Sisk and Foster 2004). We found considerable age-related differences in several HRV indices between both adolescent groups as well as between the older adolescent and both adult groups. The influence of gender on HRV akin to that observed in this study has been reported in a range of age groups in other studies (Pothineni et al. 2016; Voss et al. 2015; Abhishekh et al. 2013; Moodithaya and Avadhany 2012; Lutfi and Sukkar 2011; Reimann et al. 2010; Kuo et al. 1999). In this study, however, we did not find any gender-related differences in the group of older adolescents aged 17–20. Thus, further exploration of the effect of gender on the maturity of cardiovascular control in this age group is required.

Our findings for the influence on HRV of age factor agree with those reported for other age groups in other studies (Almeida-Santos et al. 2016; Tonhajzerova et al. 2016; Abhishekh et al. 2013; Voss et al. 2013, 2015; Boettger et al. 2010; Zhang 2007; Choi et al. 2006; Migliaro et al. 2001; Kuo et al. 1999). Changes in HRV with age have been related with dynamic changes of serum levels of testosterone, cortisol, estradiol, and adrenocorticotrophic hormones (Evans et al. 2016; Dogru et al. 2010; Fontani et al. 2004). Dynamic changes observed in the different HRV indices in the two groups of adolescents in the present study show that the autonomic regulatory influences on the cardiovascular system are actively varying during that period of life and the trend is akin to that observed in the other extreme age ranges such as fetal development (Lange et al. 2005) and elderly humans (Antelmi et al. 2004).

The influence of age and gender has been more extensively studied than the influence of heart rate on HRV. It has been long since accepted that there is a reduction in HRV with increasing heart rate and vice versa, stemming from a regulatory action of the autonomic nervous system. However, as of the 1990s, a nonlinear relationship between heart rate and its reciprocal value has been recognized of the kind we found in this study for the whole group of participants, depicted in Fig. 1. It has been recommended to use only the cardiac inter-beat duration for time and frequency domain calculations of HRV to avoid a bias (Tsuji et al. 1996; Bigger et al. 1992, 1989). Coumel et al. (1994) have emphasized that the role of heart rate in the

assessment of HRV could not be longer ignored. Those authors remark that the observed strong correlations between heart rate and HRV do not support the simplified conclusion that "looking at HRV is just a complex way to measure heart rate since the information is redundant". It has been conclusively shown that HRV indices are not a mere surrogate of heart rate (Stauss 2014). Heart rate, expressed in units of frequency or as its reciprocal (MRRi), is in fact an orthogonal factor to the time and frequency domains of calculated HRV indices before and even after applying a correction for the effect of heart rate on HRV indices (Estévez-Báez et al. 2015a, b).

The Sacha group of researchers (Sacha 2014a, b; Sacha et al. 2013; Sacha and Pluta 2008) noted the nonlinear relationships of heart rate and HRV and proposed different mathematical correcting methods. The complexity of this problem has been highlighted and the correction between the two time domains of HRV indices, SDNN and RMSSD, and heart rate has been achieved using the mathematical expressions that best fit their experimental results (Monfredi et al. 2014). In contradistinction, corrections proposed by Sacha et al. (2013) took into consideration only the variations in the length of cardiac cycle. The statistical approach to the correction of HRV due to the effect of heart rate has consisted of using the multiple linear regression models (Estévez-Báez et al. 2015a, b, 2018; Abhishekh et al. 2013; Lange et al. 2005; Antelmi et al. 2004; Tsuji et al. 1994, 1996). Recently, a simpler parsimonious mathematical correction has been proposed by van Roon et al. (2016).

In the present study, the effect of the reciprocal value of the heart rate (MRRi) on HRV was clearly described for the four age groups of healthy participants including the adolescent groups. Therefore, we strongly recommend to adjust HRV results for heart rate changes in the future studies using HRV indices. With respect to HRV changes in the younger and older groups of adolescents, we confirm the presence of an appreciable influence of the autonomic nervous system on the cardiovascular regulation, the influence that becomes progressively akin to that observed in young adults.

5 Conclusions

During adolescence, gender and particularly heart rate have a substantial influence on heart rate variability. This influence ought to be considered to avoid biases in studies on the regulatory effect of the autonomic nervous system. The observed absence of gender-related differences in heart rate variability in late adolescence of 17–20 years of age could result from the non-completely achieved maturity of the autonomic mechanisms involved with the cardiac chronotropic control, which deserves further exploration with alternative study designs. It is strongly recommended to correct for the heart rate effect on heart rate variability while evaluating the indices of heart rate variability.

Conflicts of Interest The authors declare no conflicts of interest in relation to this article.

Ethical Approval All procedures performed in studies involving human participants were in accordance with the ethical standards of the institutional and/or national research committee and with the 1964 Helsinki declaration and its later amendments or comparable ethical standards. The study was approved by the Ethics Committee of the Institute of Neurology and Neurosurgery of the Ministry of Health in Havana, Cuba.

Informed Consent Written informed consent was obtained from all individual participants included in the study.

References

Abhishekh HA, Nisarga P, Kisan R, Meghana A, Chandran S, Trichur R, Sathyaprabha TN (2013) Influence of age and gender on autonomic regulation of heart. J Clin Monit Comput 27:259–264

Almeida-Santos MA, Barreto-Filho JA, Oliveira JL, Reis FP, da Cunha Oliveira CC, Sousa AC (2016) Aging, heart rate variability and patterns of autonomic regulation of the heart. Arch Gerontol Geriatr 63:1–8

Antelmi I, de Paula RS, Shinzato AR, Peres CA, Mansur AJ, Grupi CJ (2004) Influence of age, gender, body mass index, and functional capacity on heart rate variability in a cohort of participants without heart disease. Am J Cardiol 93:381–385

Arnett JJ (2007) Emerging adulthood: what is it, and what is it good for? Child Dev Perspect 1:68–73

Bigger JT Jr, La Rovere MT, Steinman RC, Fleiss JL, Rottman JN, Rolnitzky LM, Schwartz PJ (1989) Comparison of baroreflex sensitivity and heart period variability after myocardial infarction. J Am Coll Cardiol 14:1511–1518

Bigger JT Jr, Fleiss JL, Steinman RC, Rolnitzky LM, Kleiger RE, Rottman JN (1992) Correlations among time and frequency domain measures of heart period variability two weeks after acute myocardial infarction. Am J Cardiol 69:891–898

Boettger MK, Schulz S, Berger S, Tancer M, Yeragani VK, Voss A, Bar KJ (2010) Influence of age on linear and nonlinear measures of autonomic cardiovascular modulation. Ann Noninvasive Electrocardiol 15:165–174

Bravi A, Longtin A, Seely AJ (2011) Review and classification of variability analysis techniques with clinical applications. Biomed Eng Online 10:90

Choi JB, Hong S, Nelesen R, Bardwell WA, Natarajan L, Schubert C, Dimsdale JE (2006) Age and ethnicity differences in short-term heart-rate variability. Psychosom Med 68:421–426

Christie D, Viner R (2005) Adolescent development. BMJ 330:301–304

Coumel P, Maison-Blanche P, Catuli D (1994) Heart rate and heart rate variability in normal young adults. J Cardiovasc Electrophysiol 5:899–911

Dogru MT, Basar MM, Yuvanc E, Simsek V, Sahin O (2010) The relationship between serum sex steroid levels and heart rate variability parameters in males and the effect of age. Turk Kardiyol Dern Ars 38 (7):459–465

Estévez-Báez M, Machado C, Leisman G, Brown-Martínez M, Jas-García JD, Montes-Brown J, Machado-García A, Carricarte-Naranjo C (2015a) A procedure to correct the effect of heart rate on heart rate variability indices: description and assessment. Int J Disabil Hum Dev. https://doi.org/10.1515/ijdhd-2015-0014

Estévez-Báez M, Machado C, Leisman G, Estévez-Hernández T, Arias-Morales A, Machado A, Montes-Brown J (2015b) Spectral analysis of heart rate variability. Int J Disabil Hum Dev. https://doi.org/10.1515/ijdhd-2014-0025

Estévez-Báez M, Machado C, Montes-Brown J, Jas-García J, Leisman G, Schiavi A, Machado-García A, Carricarte-Naranjo C, Carmeli E (2018) Very high frequency oscillations of heart rate variability in healthy humans and in patients with cardiovascular autonomic neuropathy. Adv Exp Med Biol 1070:49–70

Evans BE, Stam J, Huizink AC, Willemen AM, Westenberg PM, Branje S, Meeus W, Koot HM, van Lier PA (2016) Neuroticism and extraversion in relation to physiological stress reactivity during adolescence. Biol Psychol 117:67–79

Fontani G, Lodi L, Felici A, Corradeschi F, Lupo C (2004) Attentional, emotional and hormonal data in participants of different ages. Eur J Appl Physiol 92:452–461

Goldberger JJ, Johnson NP, Haris Subacius H, Jason Ng J, Greenland P (2014) Comparison of the physiological and prognostic implications of the heart rate versus the RR interval. Heart Rhythm 11:1925–1933

Kuo TB, Lin T, Yang CC, Li CL, Chen CF, Chou P (1999) Effect of aging on gender differences in neural control of heart rate. Am J Phys 277:H2233–H2239

Kuusela T (2013) Methodological aspects of heart rate variability analysis. heart rate variability (HRV) signal analysis: clinical applications, vol I, 1st edn. CRC Press Taylor & Francis Group, Boca Raton/London/New York

Lange S, Van Leeuwen P, Geue D, Hatzmann W, Gronemeyer D (2005) Influence of gestational age, heart rate, gender and time of day on fetal heart rate variability. Med Biol Eng Comput 43:481–486

Lutfi MF, Sukkar MY (2011) The effect of gender on heart rate variability in asthmatic and normal healthy adults. Int J Health Sci 5:146–154

Machado C, Estevez M, Rodriguez R, Perez-Nellar J, Chinchilla M, DeFina P, Leisman G, Carrick FR, Melillo R, Schiavi A, Gutierrez J, Carballo M, Machado A, Olivares A, Perez-Cruz N (2014) Zolpidem arousing effect in persistent vegetative state patients: autonomic, EEG and behavioral assessment. Curr Pharm Des 20:4185–4202

Machado-Ferrer Y, Estevez M, Machado C, Hernandez-Cruz A, Carrick FR, Leisman G, Melillo R, Defina P, Chinchilla M, Machado Y (2013) Heart rate variability for assessing comatose patients with different Glasgow Coma Scale scores. Clin Neurophysiol 124:589–597

Mestanikova A, Ondrejka I, Mestanik M, Hrtanek I, Snircova E, Tonhajzerova I (2016) Electrodermal activity in adolescent depression. Adv Exp Med Biol 935:83–88

Michels N, Clays E, De Buyzere M, Huybrechts I, Marild S, Vanaelst B, De Henauw S, Sioen I (2013) Determinants and reference values of short-term heart rate variability in children. Eur J Appl Physiol 113:1477–1488

Migliaro ER, Contreras P, Bech S, Etxagibel A, Castro M, Ricca R, Vicente K (2001) Relative influence of age, resting heart rate and sedentary life style in short-term analysis of heart rate variability. Braz J Med Biol Res 34:493–500

Monfredi O, Lyashkov AE, Johnsen AB, Inada S, Schneider H, Wang R, Nirmalan M, Wisloff U, Maltsev VA, Lakatta EG, Zhang H, Boyett MR (2014) Biophysical characterization of the underappreciated and important relationship between heart rate variability and heart rate. Hypertension 64 (6):1334–1343

Moodithaya S, Avadhany ST (2012) Gender differences in age-related changes in cardiac autonomic nervous function. J Aging Res 2012:679345

Nieminen T, Kahonen M, Koobi T, Nikus K, Viik J (2007) Heart rate variability is dependent on the level of heart rate. Am Heart J 154:e13

Pothineni NV, Shirazi LF, Mehta JL (2016) Gender differences in autonomic control of the cardiovascular system. Curr Pharm Des 22:3829–3834

Reimann M, Friedrich C, Gasch J, Reichmann H, Rudiger H, Ziemssen T (2010) Trigonometric regressive spectral analysis reliably maps dynamic changes in baroreflex sensitivity and autonomic tone: the effect of gender and age. PLoS One 5:e12187

Sacha J (2013) Why should one normalize heart rate variability with respect to average heart rate. Front Physiol 4:306

Sacha J (2014a) Interaction between heart rate and heart rate variability. Ann Noninvasive Electrocardiol 19 (3):207–216

Sacha J (2014b) Interplay between heart rate and its variability: a prognostic game. Front Physiol 5:347

Sacha J, Pluta W (2008) Alterations of an average heart rate change heart rate variability due to mathematical reasons. Int J Cardiol 128:444–447

Sacha J, Barabach S, Statkiewicz-Barabach G, Sacha K, Muller A, Piskorski J, Barthel P, Schmidt G (2013) How to strengthen or weaken the HRV dependence on heart rate—description of the method and its perspectives. Int J Cardiol 168:1660–1663

Sassi R, Cerutti S, Lombardi F, Malik M, Huikuri HV, Peng CK, Schmidt G, Yamamoto Y, Document R, Gorenek B, Lip GY, Grassi G, Kudaiberdieva G, Fisher JP, Zabel M, Macfadyen R (2015) Advances in heart rate variability signal analysis: joint position statement by the e-Cardiology ESC Working Group and the European Heart Rhythm Association co-endorsed by the Asia Pacific Heart Rhythm Society. Europace 17:1341–1353

Sharma VK, Subramanian SK, Arunachalam V, Rajendran R (2015) Heart rate variability in adolescents - normative data stratified by sex and physical activity. J Clin Diagn Res 9:CC08–CC13

Sisk CL, Foster DL (2004) The neural basis of puberty and adolescence. Nat Neurosci 7:1040–1047

Sosnowski M (2011) Heart rate variability. In: Macfarlane PW, van Oosterom A, Pahlm O, Kligfield PM, Janse MJC (eds) Comprehensive electrocardiology. Springer-Verlag Limited, London

Stauss HM (2014) Heart rate variability: just a surrogate for mean heart rate? Hypertension 64(6):1184–1186

Task Force of ESC and NASPE (1996) Heart rate variability. Standards of measurement, physiological interpretation, and clinical use. Task Force of the European Society of Cardiology and the North American Society of Pacing and Electrophysiology. Eur Heart J 17:354–381

Tonhajzerova I, Visnovcova Z, Mestanikova A, Jurko A, Mestanik M (2016) Cardiac vagal control and depressive symptoms in response to negative emotional stress. Adv Exp Med Biol 934:23–30

Tsuji H, Venditti FJJ, Manders ES, Evans JC, Larson MG, Feldman CL, Levy D (1994) Reduced heart rate variability and mortality risk in an elderly cohort. The Framingham Heart Study. Circulation 90:878–883

Tsuji H, Larson MG, Venditti FJ Jr, Manders ES, Evans JC, Feldman CL, Levy D (1996) Impact of reduced heart rate variability on risk for cardiac events. The Framingham Heart Study. Circulation 94:2850–2855

van Roon AM, Snieder H, Lefrandt JD, de Geus EJ, Riese H (2016) Parsimonious correction of heart rate variability for its dependency on heart rate. Hypertension 68(5):e63–e65

Voss A, Schroeder R, Fischer C, Heitmann A, Peters A, Perz S (2013) Influence of age and gender on complexity measures for short term heart rate variability analysis in healthy participants. Conference proceedings: annual international conference of the IEEE engineering in medicine and biology society ieee engineering in medicine and biology society conference 2013, pp 5574–5577

Voss A, Schroeder R, Heitmann A, Peters A, Perz S (2015) Short-term heart rate variability--influence of gender and age in healthy participants. PLoS One 10: e0118308

Zhang J (2007) Effect of age and sex on heart rate variability in healthy participants. J Manip Physiol Ther 30:374–379

Adv Exp Med Biol - Clinical and Experimental Biomedicine (2019) 4: 35–40
https://doi.org/10.1007/5584_2018_328
© Springer Nature Switzerland AG 2019
Published online: 31 January 2019

Hand-Foot Syndrome and Progression-Free Survival in Patients Treated with Sunitinib for Metastatic Clear Cell Renal Cell Carcinoma

Jakub Kucharz, Monika Budnik, Paulina Dumnicka, Maciej Pastuszczak, Beata Kuśnierz-Cabala, Tomasz Demkow, Katarzyna Popko, and Pawel Wiechno

Abstract

Patients with metastatic clear cell renal cell carcinoma (mRCC) typically receive systemic treatment with tyrosine kinase inhibitors (TKI). Side effects include the hand-foot syndrome (HFS), tiredness, nausea, decreased appetite, diarrhea, myelosuppression, and hypertension. This study seeks to define the relationship between the incidence of HFS after the first cycle of treatment with sunitinib as the first-line treatment for mRCC (50 mg/day, 6-week schedule: 4 weeks on and 2 weeks off) and progression-free survival. We found that patients, treated with sunitinib for mRCC, who did not experience HFS had the median progression-free survival of 9.8 months. HFS symptoms appeared in 20% of patients after the first treatment cycle. The appearance of HFS was a predictor of a longer progression-free survival. In fact, progression-free survival was elongated in the HFS group over and beyond the observation period of 60 months, which rendered the median progression-free survival calculation impossible. These findings reaffirm the importance of monitoring skin toxicity during treatment with TKI. We conclude that the appearance of adverse skin symptoms presages better outcomes in patients treated with sunitinib for mRCC.

J. Kucharz, T. Demkow, and P. Wiechno
Department of Uro-Oncology, Maria Sklodowska-Curie Memorial Cancer Center and Institute of Oncology, Warsaw, Poland

M. Budnik (✉)
First Chair and Department of Cardiology, Warsaw Medical University, Warsaw, Poland
e-mail: moni.budnik@gmail.com

P. Dumnicka
Department of Medical Diagnostics, Jagiellonian University Medical College, Cracow, Poland

M. Pastuszczak
Department of Dermatology, Jagiellonian University Medical College, Cracow, Poland

B. Kuśnierz-Cabala
Department of Clinical Biochemistry, Jagiellonian University Medical College, Cracow, Poland

K. Popko
Department of Laboratory Medicine and Clinical Immunology of Developmental Age, Warsaw Medical University, Warsaw, Poland

Keywords

Hand-foot syndrome · Renal cell carcinoma · Sunitinib · Survival · Tyrosine kinase inhibitors

1   Introduction

Renal cell carcinoma accounts for 3% and 5% of malignancies in women and men, respectively (Escudier et al. 2016). The onset is most frequent in the sixth decade of life. When the carcinoma is limited to the kidney, it is typically treated with surgery, while metastatic clear cell renal cell (mRCC) patients require individualized treatment. Some patient may benefit from metastasectomy that should be considered when mRCC is initially diagnosed. When surgical treatment is impossible, patients qualify for systemic treatment. Currently, systemic treatment protocols for mRCC include tyrosine kinase inhibitors (TKI) or drugs that inhibit the mammalian target of rapamycin (mTOR), a serine/threonine-specific protein kinase. The first-line treatment is selected on the basis of Memorial Sloan-Kettering Cancer Center (MSKCC) risk criteria (Motzer et al. 1999). According to the European Society for Medical Oncology (ESMO) guidelines, patients in low- and intermediate-risk groups receive bevacizumab and interferon-alpha (IFN-α) or TKI, either pazopanib or sunitinib (Escudier et al. 2016). These agents inhibit tyrosine kinase receptors, which participate in tumor growth, angiogenesis, and metastasis. Sunitinib is an inhibitor of vascular endothelial growth factor receptors (VEGFR-1, VEGFR-2, and VEGFR-3), stem cell factor (SCF) that binds and activates the receptor tyrosine kinase c-Kit, platelet-derived growth factor receptors (PDGFRα and β), Fms-like tyrosine kinase 3 (FLT3), colony-stimulating factor-1 receptors (CSF-1R), and glial cell line-derived neurotrophic factor receptors (RET) (Chow and Eckhardt 2007). In a sunitinib approval study that compared the effectiveness of sunitinib and IFN-α in patients with metastatic clear cell renal cell carcinoma, median progression-free survival was 11 months for patients treated with sunitinib and 5 months for those treated with IFN-α (Motzer et al. 2007). Treatment is usually well tolerated. The most frequent side effects include the hand-foot syndrome (HFS), tiredness, nausea, loss of appetite, diarrhea, myelosuppression, and hypertension. Some studies have reported an association between the incidence of adverse effects and treatment outcomes, including progression-free survival, overall response rate, and overall survival (Maráz et al. 2018; Buda–Nowak et al. 2017; Kucharz et al. 2016; Nakano et al. 2013; Ravaud and Schmidinger 2013; Poprach et al. 2012; Bono et al. 2011; Schmidinger et al. 2011).

The HFS, also known as palmoplantar erythrodysesthesia, is a clinical manifestation of dermal toxicity of many oncological drugs. It encompasses skin reddening, blisters, hyperkeratotic changes, and associated pain, as well as paresthesias and dysesthesias (Lipworth et al. 2009; Chu et al. 2008). These symptoms typically occur on the palm of the hand and the sole of the foot. The goal of this study was to evaluate the association between the appearance in the first treatment cycle of HFS and progression-free survival in patients with mRCC treated with sunitinib.

2   Methods

The study encompassed 28 patients with clear cell mRCC treated with sunitinib during 2014–2017. Patients were selected according to the following criteria:

- ≥18 years of age
- Histopathological diagnosis of clear cell RCC
- Metastases at the time of the initial diagnosis or a relapse after radical therapy
- Being after nephrectomy
- Classified either to low or intermediate MSKCC risk group (0–2 points)
- Systemic sunitinib as the first-line treatment
- Completeness of medical files

Patients who had non-RCC-related malignant tumors were excluded from the study. Clinical data were collected, including demographic, pathological, laboratory, and radiological indices. The incidence and intensity of HFS were assessed after the completion of the first cycle of treatment, i.e., on the 43rd day of treatment, on the basis of the Common Terminology Criteria for Adverse Events (CTC-AE) scale, version 3.0 (Janusch et al. 2006). The intensity of HFS is shown in Table 1.

Continuous variables were expressed as medians with lower and upper quartiles and nominal variables as the number or percentage of patients. Patient groups were compared using the Mann-Whitney U and chi-squared tests. The starting point for progression-free survival was the beginning of treatment, and the ending point was either disease progression or death. We used the Kaplan-Meier estimates to compute progression-free survival. Data of patients who did not experience either disease progression or death at the conclusion of the study were censored. The correlation between the HFS incidence and progression-free survival was evaluated with the Cox proportional hazards regression and the log-rank test. Differences were considered statistically significant at $p < 0.05$. All statistical calculations were conducted using a commercial STATISTICA package v12.0 (StatSoft, Tulsa, OK).

3   Results

Table 2 shows clinical characteristics of patients. In 6 (21%) out of the 28 patients, there appeared HFS of grades 1 to 3 after the first cycle of sunitinib treatment. The initial diagnosis of a metastatic clear cell renal cell carcinoma was more frequent among patients who did not develop HFS symptoms than in those who did; the difference was statistically significant. The HFS and non-HFS patient groups did not differ in terms of age, gender, weight and BMI, grade of renal cell carcinoma malignancy, time from onset of systemic treatment, or the level of pre-treatment functioning.

We found a significant difference between the progression-free survival time in patients who experienced HFS symptoms after the first cycle of sunitinib treatment and those who did not. The median progression-free survival amounted to 9.8 months in patients without HFS, but it extended much over that time in patients with HFS, which rendered it incalculable during the study observation period of 60 months (Fig. 1).

Table 1 Grading of the hand-foot syndrome (HFS)

Grade	Description
G1	Mild skin changes: redness, swelling, and hyperkeratosis; no pain
G2	Moderate skin changes: redness, swelling hyperkeratosis, scaling, blisters; pain paresthesias; interfering with fine motor movement
G3	Severe skin changes: redness, swelling, hyperkeratosis, scaling, blisters; pain, paresthesias; interfering with daily activities and self-care

Table 2 Clinical characteristics of patients

Variable	HFS (n = 6)	No HFS (n = 22)	p
Age (range): years	60 (46–65)	66 (63–69)	0.20
Women/men: n (%)	3 (50)/3 (50)	6 (27)/16 (73)	0.30
Weight (range): kg	81 (75–88)	72 (65–77)	0.08
BMI (range): kg/m^2	29 (27–30)	27 (25–29)	0.09
Metastatic disease on the initial diagnosis: n (%)	3 (50)	20 (91)	0.02
Fuhrman nuclear grades 1–2/3–4 according to the American Urological Association (2018): n (%)	2 (33)/4 (67)	8 (36)/14 (64)	0.90
Time from diagnosis to onset of systemic treatment <1 year: n (%)	2 (33)	13 (59)	0.30
MSKCC prognosis: good/intermediate – n (%)	4 (67)/2 (33)	4 (18)/18 (82)	0.02
Performance status before treatment with sunitinib, according to the ECOG (2018) scale: 0/1 – n (%)	3 (50)/3 (50)	10 (45)/12 (55)	0.90

BMI body mass index, *MSKCC* Memorial Sloan-Kettering Cancer Center risk criteria, *ECOG* Eastern Cooperative Oncology Group – Performance Status

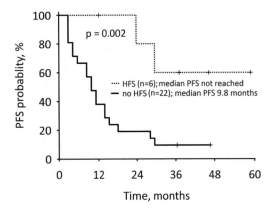

Fig. 1 Progression-free survival (PFS) in patients who did experience HFS after the first cycle of treatment (dotted line) and those who did not (continuous line). The plot shows log-rank test p-value

The association between the appearance of HFS and progression-free survival was also significant in the simple Cox regression analysis (Table 3; Model 1). The multiple Cox regression adjusted for the variables that differed significantly between HFS and non-HFS groups confirmed that HFS significantly predicted progression-free survival independently of the patients' MSKCC prognosis category (Table 3; Model 2) and of renal cancer metastases found upon the initial diagnosis (Table 3; Model 3).

Correlation between the incidence of HFS after the first cycle of treatment and progression-free survival was independent of the patients' MSKCC prognostic category or the presence of metastases on the initial diagnosis (Table 3).

4   Discussion

The HFS symptoms appear in the course of systemic treatment with classical cytostatic agents such as 5-fluorouracil (Janusch et al. 2006), capecitabine, pegylated liposomal doxorubicin (Lorusso et al. 2007), or the targeted tyrosine kinase inhibitors such as sunitinib or sorafenib (Motzer et al. 2007, Escudier et al. 2007). A wide spectrum of HFS symptoms ranges from redness that causes no discomfort to complications that prevent the patient from performing basic functions and require opioid

treatment to alleviate pain. The pathogenesis of HFS is unclear and probably differs for different types of drugs. For cytostatics, such as pegylated liposomal doxorubicin, the pathogenesis is linked to excretion of the agent from the eccrine glands that are highly concentrated in the palmar and plantar skin. The underlying mechanism also has to do with micro-abrasions caused by daily activities (Lorusso et al. 2007).

When HFS is induced by administration of sunitinib and by other tyrosine kinase inhibitors, its mechanism seems linked to these agents' influence on the endothelial and fibroblastic cells that play an essential role in tissue repair (Li and Gu 2017; Wyganowska–Swiatkowska et al. 2016). A new theory unique to the tyrosine kinase inhibitors explains the appearance of HFS as a result of inhibition of receptors for both platelet-derived growth factor (PDGF-R) and vascular endothelial growth factor (VEGF-R), which hinders the vascular repair mechanisms (Pożarowska and Pożarowski 2016). This disturbance would be the most clinically apparent in the areas exposed to high pressure and repeated trauma, such as palms and soles (Lipworth et al. 2009). Further, the skin rash of HFS could be used as a surrogate marker for drug activity. This presumption made it rational for us to evaluate in the present study the extent of any relationship between the occurrence of HFS and progression-free survival in patients with mRCC who were treated with the tyrosine kinase inhibitor sunitinib. We found that HFS symptoms appeared after the first treatment cycle in 20% of patients, and the appearance of HFS was a predictor of a longer progression-free survival. In fact, progression-free time was elongated in the HFS group over and beyond the observation period of 60 months, which rendered the median progression-free survival calculation impossible (Fig. 1). At the end of that period, the proportion of patients with progression-free survival amounted to about 60%. Our findings were, in general, in line with those of other studies that show a variable incidence of all grades of HFS in mRCC in a range of 14–41%. Poprach et al. (2012) have investigated the relationship between skin toxicity and treatment outcomes in 705 patients receiving sunitinib as the first-line treatment for

Table 3 Hand-foot syndrome as a predictor of progression-free survival in univariate and multivariate Cox regression analysis

Independent variable	Relative risk of disease progression (95% confidence interval)		
	Model 1	Model 2	Model 3
HFS	0.16 (0.04–0.74)	0.19 (0.04–0.85)	0.17 (0.03–0.86)
	p = 0.02	p = 0.03	p = 0.03
Intermediate MSKCC prognosis		1.51 (0.53–4.28)	
		p = 0.40	
Metastatic disease on initial diagnosis			0.98 (0.20–4.98)
			p = 0.90

HFS hand-foot syndrome, *MSKCC* Memorial Sloan-Kettering Cancer Center risk criteria

mRCC. In that study, skin toxicity has been defined as either HFS (any grade) or rash (grade 3 or 4), and patients who experienced these adverse events show a better treatment outcome. The median progression-free survival for patients with significant skin toxicity was 20.8 months, while that for patients without skin toxicity was 11.1 months. Likewise, median overall survival amounted to 43.0 and 31.0 months in the respective groups. It is important to note, however, that some of the patients in that study had received cytokine treatment before the study, and the first-line treatment with TKI was in fact their second-line treatment for mRCC. Michaelson et al. (2011) have reported the HFS prevalence at 23% in mRCC patients treated with sunitinib. They investigated a large cohort of 770 patients who had been given two treatment protocols: 71% of patients received sunitinib in a 6-week treatment cycle (50 mg/day, 4 weeks on, 2 weeks off), while the remaining patients received continuous administration of sunitinib (37.5 mg/day). The overall response rate was 44.6% and 32.7%, respectively. Also in that study, patients who developed HFS had significantly better treatment outcomes than those who did not, with median progression-free survival of 14.3 and 8.3 months and the overall survival of 38.3 and 18.0 months respectively. The interpretation of those findings is, however, hampered, as the second option of continuous administration of sunitinib mentioned above is currently considered a suboptimal treatment.

In conclusion, the present study reaffirms the importance of monitoring skin toxicity during treatment with TKI. The appearance of adverse skin symptoms, which most often arise in 30 odd days of treatment, i.e., around the completion of the first treatment cycle, speaks for a greater efficacy of anticancer treatment. Thus, skin symptoms are liable to be a presage of a better outcome in patients treated with sunitinib for mRCC.

Conflicts of Interest JK received a research grant from Novartis Poland and speaker honoraria from Pfizer, Bayer, and IPSEN companies. PW received speaker honoraria from Pfizer, Bayer, Novartis, and IPSEN. The remaining authors declare no conflicts of interest in relation to this article.

Ethical Approval All procedures performed in studies involving human participants were in accordance with the ethical standards of the institutional and/or national research committee and with the 1964 Helsinki Declaration and its later amendments or comparable ethical standards. The study was approved by the Bioethics Committee (permission no. 38/2018) of the Maria Sklodowska-Curie Memorial Cancer Center and Institute of Oncology in Warsaw, Poland.

Informed Consent Informed consent was obtained from all individual participants included in the study.

References

American Urological Association (2018) Clear cell renal cell carcinoma: Fuhrman nuclear grade. https://www.auanet.org/education/auauniversity/education-products-and-resources/pathology-for-urologists/kidney/renal-cell-carcinomas/clear-cell-renal-cell-carcinoma-fuhrman-nuclear-grade. Accessed on 30 Nov 2018

Bono P, Rautiola J, Utriainen T, Joensuu H (2011) Hypertension as predictor of sunitinib treatment outcome in metastatic renal cell carcinoma. Acta Oncol 50:569–573

Buda–Nowak A, Kucharz J, Dumnicka P, Kuzniewski M, Herman RM, Zygulska AL, Kusnierz–Cabala B (2017) Sunitinib–induced hypothyroidism predicts progression–free survival in metastatic renal cell carcinoma patients. Med Oncol 34(4):68

Chow LQ, Eckhardt SG (2007) Sunitinib: from rational design to clinical efficacy. J Clin Oncol 25:884–896

Chu D, Lacouture ME, Fillos T, Wu S (2008) Risk of hand–foot skin reaction with sorafenib: a systematic review and meta–analysis. Acta Oncol 47(2):176–186

ECOG (2018) Eastern cooperative oncology group. Performance status. https://www.mdcalc.com/eastern-cooperative-oncology-group-ecog-performance-status. Accessed on 30 Nov 2018

Escudier B, Eisen T, Stadler W, Szczylik C, Oudard S, Siebels M, Negrier S, Chevreau C, Solska E, Desai A, Rolland F, Demkow T, Hutson T, Gore M, Freeman S, Schwartz B, Shan M, Simantov R, Bukowski R, for the TARGET Study Group (2007) Sorafenib in advanced clear–cell renal–cell carcinoma. N Engl J Med 356 (2):125–134

Escudier B, Porta C, Schmidinger M, Rioux–Leclercq N, Bex A, Khoo V, Gruenvald V, Horwich A, ESMO Guidelines Committee (2016) Renal cell carcinoma: ESMO Clinical Practice Guidelines for diagnosis, treatment and follow–up. Ann Oncol 27(suppl 5):58–68

Janusch M, Fischer M, Marsch WC, Holzhausen HJ, Kegel T, Helmbold P (2006) The hand–foot syndrome – a frequent secondary manifestation in antineoplastic chemotherapy. Eur J Dermatol 16(5):494–499

Kucharz J, Giza A, Dumnicka P, Kuzniewski M, Kusnierz–Cabala B, Bryniarski P, Herman R, Zygulska AL, Krzemieniecki K (2016) Macrocytosis during sunitinib treatment predicts progression–free survival in patients with metastatic renal cell carcinoma. Med Oncol 33(10):109

Li J, Gu J (2017) Hand–foot skin reaction with vascular endothelial growth factor receptor tyrosine kinase inhibitors in cancer patients: a systematic review and meta–analysis. Crit Rev Oncol Hematol 119:50–58

Lipworth AD, Robert C, Zhu AX (2009) Hand–foot syndrome (hand–foot skin reaction, palmar–plantar erythrodysesthesia): focus on sorafenib and sunitinib. Oncology 77(5):257–271

Lorusso D, Di Stefano A, Carone V, Fagotti A, Pisconti S, Scambia G (2007) Pegylated liposomal doxorubicin-related palmar–plantar erythrodysesthesia ('hand–foot' syndrome). Ann Oncol 18(7):1159–1164

Maráz A, Cserháti A, Uhercsák G, Szilágyi É, Varga Z, Révész J, Kószó R, Varga L, Kahán Z (2018) Dose escalation can maximize therapeutic potential of sunitinib in patients with metastatic renal cell carcinoma. BMC Cancer 18(1):296

Michaelson MD, Cohen DP, Li S, Motzer RJ, Escudier B, Barrios CH, Burnett PE, Puzanov I (2011) Handfoot syndrome (HFS) as a potential biomarker of efficacy in patients (pts) with metastatic renal cell carcinoma (mRCC) treated with sunitinib (SU). J Clin Oncol 29 (Suppl 7):Abstr 320

Motzer RJ, Mazumdar M, Bacik J, Berg W, Amsterdam A, Ferrara J (1999) Survival and prognostic stratification of 670 patients with advanced renal cell carcinoma. J Clin Oncol 17(8):2530–2540

Motzer RJ, Hutson TE, Tomczak P, Michaelson MD, Bukowski RM, Rixe O, Oudard S, Negrier S, Szczylik C, Kim ST, Chen I, Bycott PW, Baum CM, Figlin RA (2007) Sunitinib versus interferon alfa in metastatic renal–cell carcinoma. N Engl J Med 356:115–124

Nakano K, Komatsu K, Kubo T, Natsui S, Nukui A, Kurokawa S, Kobayashi M, Morita T (2013) Hand–foot skin reaction is associated with the clinical outcome in patients with metastatic renal cell carcinoma treated with sorafenib. Jpn J Clin Oncol 43 (10):1023–1029

Poprach A, Pavlik T, Melichar B, Puzanov I, Dusek L, Bortlicek Z, Vyzula R, Abrahamova J, Buchler T, Czech Renal Cancer Cooperative Group (2012) Skin toxicity and efficacy of sunitinib and sorafenib in metastatic renal cell carcinoma: a national registry–based study. Ann Oncol 23(12):3137–3143

Pożarowska D, Pożarowski P (2016) The era of anti–vascular endothelial growth factor (VEGF) drugs in ophthalmology, VEGF and anti–VEGF therapy. Cent Eur J Immunol 41:311–316

Ravaud A, Schmidinger M (2013) Clinical biomarkers of response in advanced renal cell carcinoma. Ann Oncol 24:2935–2942

Schmidinger M, Vogl UM, Bojic M, Lamm W, Heinzl H, Haitel A, Clodi M, Kramer G, Zielinski CC (2011) Hypothyroidism in patients with renal cell carcinoma: blessing or curse. Cancer 117(3):534–544

Wyganowska–Swiatkowska M, Urbaniak P, Szkaradkiewicz A, Jankun J, Kotwicka M (2016) Effects of chlorhexidine, essential oils and herbal medicines (salvia, chamomile, calendula) on human fibroblast in vitro. Cent Eur J Immunol 41:125–131

Adv Exp Med Biol - Clinical and Experimental Biomedicine (2019) 4: 41–48
https://doi.org/10.1007/5584_2018_303
© Springer Nature Switzerland AG 2018
Published online: 16 November 2018

Elastography in the Diagnosis of Pancreatic Malignancies

Przemysław Dyrla, Jerzy Gil, Stanisław Niemczyk,
Marek Saracyn, Krzysztof Kosik, Sebastian Czarkowski,
and Arkadiusz Lubas

Abstract

The study aimed to determine the usefulness of the elastography in the diagnosis of malignancy of solid pancreatic tumors. There were 123 patients (F/M; 51/72, aged 62 ± 14) enrolled into the study with the diagnosis of pancreatic masses. Malignant pancreatic adenocarcinoma was identified in 78 patients and an inflammatory mass corresponding to chronic pancreatitis in the remaining 45 patients. The mass elasticity of a tumor (A–elasticity) and a reference zone (B–elasticity) and the B/A strain ratio were measured. All these elastographic parameters differed between groups and correlated significantly with malignancies ($r = 0.841$; $r = -0.834$; $r = 0.487$, respectively). Receiver operating characteristic (ROC) analysis showed that A–elasticity between 0.05% and 0.14% alone, as well as the B/A strain ratio between 7.87 and 18.23 alone, enabled the recognition of all malignant pancreatic tumors with 100% sensitivity and $\geq 97.8\%$ specificity. Surprisingly, B–elasticity alone also was helpful in recognizing malignant tumors (71% sensitivity, 80% specificity, 0.74 accuracy, and 0.792 area under the curve), although it appeared worse than A–elasticity and B/A strain ratio ($p < 0.001$). In multivariable regression analysis, A–elasticity identified 89.5% of malignancies ($p < 0.001$). A–elasticity and B–elasticity were the only significant independent factors influencing the tumor identification ($r^2 = 0.927$; $p < 0.001$). The assessment of tumor elasticity appears sufficient to identify malignant tumors of the pancreas.

P. Dyrla, J. Gil, and K. Kosik
Department of Gastroenterology, Military Institute of Medicine, Warsaw, Poland

S. Niemczyk (✉) and A. Lubas
Department of Internal Medicine, Nephrology and Dialysis, Military Institute of Medicine, Warsaw, Poland
e-mail: sniemczyk@wim.mil.pl

M. Saracyn
Department of Endocrinology and Isotope Therapy, Military Institute of Medicine, Warsaw, Poland

S. Czarkowski
Department of Radiology, Military Institute of Medicine, Warsaw, Poland

Keywords

Elastography · Endosonography · Pancreatic malignancy · ROC analysis · Strain ratio · Tumor elasticity

1 Introduction

Focal pancreatic masses pose a serious diagnostic problem. The differentiation between malignant and benign tumors and the evaluation of the possibility of surgical treatment are the most important in the diagnosis of solid pancreatic tumors. Among malignant tumors, one can distinguish adenocarcinoma, neuroendocrine tumors, solid pseudopapillary neoplasms, stromal tumors, lymphomas, and metastases. Adenocarcinoma is the worst prognosing pancreatic neoplasm, for which the average survival time is less than 6 months and a 5-year survival applies to less than 5% of patients (Theoharis 2008). It is the most frequent primary malignant tumor of the exocrine part of the pancreas, constituting approximately 90–95% of cases. Typically, it is located in the head of the pancreas (60%), less frequently in the body (15%), and the least in the tail (5%) (Cascinu et al. 2010). Twenty percent of tumors are of disseminated character. Less than 30% of pancreatic cancers are eligible for surgery at diagnosis, since of early lymphatic and hematogenous metastases (Seufferlein et al. 2012). A second group of pancreatic neoplasms are tumors derived from pancreatic islets. These are relatively rare neoplasms (1 *per* 100 thousand population) of neuroendocrine origin, representing 1–5% of pancreatic tumors (Davis et al. 2009). The pancreas is a place where B-cell non-Hodgkin lymphoma can be localized. The original location of these cells in the pancreas is infrequent and associated with immunosuppression (Aslam and Yee 2006). The most common metastasis in the pancreas comes from kidney cancer, but bronchogenic carcinoma, breast cancer, colon neoplasms, melanoma, and soft tissue tumor metastases can be found as well (Aslam and Yee 2006; Scatarige et al. 2001; Merkle et al. 2000). Imaging methods play the most important role in the diagnostic process of solid pancreatic tumors from the moment of detection and help evaluate the disease severity and plan the treatment. There is no perfect, widely accepted method that would meet the challenge of the assessment of pancreatic malignancy and the possible surgical approach to it. Currently, endosonography is the best method for detecting solid pancreatic masses and, combined with tissue elastography, enables the determination of a nature of solid pancreatic tumors. The present study seeks to define the usefulness of the elastography parameters alone in the diagnosis of pancreatic malignancy.

2   Methods

The investigation was designed as a prospective single-center study conducted in the Department of Gastroenterology of the Military Institute of Medicine in Warsaw, Poland. There were 123 patients (F/M; 51/72; mean age 62 ± 14 years), with the diagnosis of pancreatic solid masses, included in this 2-year follow-up study. The inclusion criterion was a solid pancreatic mass diagnosed with imaging techniques. Pancreatic tumors with cystic or liquid components were excluded. Each patient was subjected to endosonography (EUS) and computed tomography (CT). Elastography images were recorded using a Pentax EG-3870 UTK EUS linear probes (Pentax, Tokyo, Japan) combined with a Hitachi Preirus ultrasound machine (Hitachi Medical Systems, Tokyo, Japan). For the estimation of mass elasticity, two different regions of interest (ROI) were selected. In the tumor area, A–elasticity was assessed, whereas in a soft peri-pancreatic reference zone outside the tumor, B–elasticity was assessed. The measurement was made two times in each ROI, and the mean of the two was considered for statistics. The B/A strain ratio of tissue elasticity was calculated as proposed by Iglesias–Garcia et al. (2010).

Data were reported as means $\pm$SD. Differences between malignant and benign tumor groups were evaluated using Student's t-test or Mann–Whitney U test for normal and non-normal data distribution, respectively. Spearman's correlation analysis was used to determine the relationship between the elastographic and other variables. The diagnostic accuracy and the predictive value of elastographic variables for the identification of pancreatic malignancy were assessed using the receiver operating curve (ROC) analysis. A stepwise multivariable linear regression analysis was used to determine factors independently connected with the occurrence of malignancy. A p-value <0.05 defined statistically significant differences. The analysis was performed with a commercial statistical package of Statistica v.12 (StatSoft Inc., Tulsa, Oklahoma, USA).

3   Results

EUS elastography and EUS fine needle aspiration (EUS–FNA) were performed in all 123 patients. No complication occurred during the procedures. The mean maximum diameter of pancreatic masses was 3.77 ± 1.33 cm. Overall, women were older than men (59 ± 14 vs. 66 ± 13 years, respectively; p = 0.005) in the investigated population. Pancreatic masses were located in the pancreas head in 87 patients, pancreas body in 28 patients, and in the pancreas tail in 8 patients. Cytologic diagnosis was obtained by EUS–FNA in all patients after a mean of two needle passes. A malignant pancreatic adenocarcinoma (Group 1) was identified in 78 patients (39 women aged 69 ± 12, and 39 men aged 64 ± 13; p = 0.03). It was diagnosed by EUS–FNA cytology in 54 cases and by histology of surgical specimens in 24 patients who underwent surgery. The diagnosis of an inflammatory mass in the context of chronic pancreatitis (Group 2; n = 45) was based on the presence of inflammatory cells in cytology, along with the EUS criteria for chronic pancreatitis and with the CT images consistent with the disease (Perez–Johnston et al. 2012; Shimosegawa et al. 2011; Catalano et al. 2009; Wiersema et al. 1993). The diagnosis of inflammation was confirmed by histology of surgical specimens in five patients who underwent surgery. After a follow-up period of about 6–8 months, malignancy was definitely excluded in the remaining 40 patients of Group 2 by clinical and laboratory evaluation, consisting of CT, EUS, and EUS-guided FNA. Differences in age and gender between patients with pancreatic malignant (Group 1) and non-malignant (Group 2) tumors are presented in Table 1.

The frequency of occurrence of pancreatic malignancies, in relation to all recognized tumors, was similar in all pancreatic regions (p = 0.459). The number of estimated elastographic parameters in tumors (A–elasticity) and in referential places (B–elasticity) differed between groups (Table 2). Considering all 123 patients investigated, A–elasticity correlated somehow stronger with malignancy than the B/A strain

Table 1 Demographic features of patients. Group 1 – malignant tumor (pancreatic adenocarcinoma) and Group 2 – benign tumor (chronic pancreatitis)

	Group 1 (n = 78)	Group 2 (n = 45)	p-value
F/M	39/39	12/33	
Age (F) (yr)	69.2 ± 11.8*	56.8 ± 13.5	0.007
Age (M) (yr)	63.7 ± 13.0*	53.8 ± 13.5	0.002

F female, *M* male
*p = 0.03 for the inter-gender difference in Group 1

ratio did (Table 3). In ROC analysis, either the value of A–elasticity alone between 0.05% and 0.14% or the value of B/A strain ratio alone between 7.87 and 18.23 enabled the almost perfect recognition of all malignant pancreatic tumors (100% sensitivity, 97.8–100% specificity, 0.992–1.0 accuracy, 1.0 AUC, positive predicted value (PPV) of 0.987–1.0, and negative predicted value (NPV) of 1.0) (Figs. 1, 2, and 3).

In comparison to the parameters above outlined, B–elasticity came out significantly worse (p < 0.001) for the identification of malignant tumors (71% sensitivity, 80% specificity, 0.74 accuracy, 0.792 AUC, PPV of 0.859, and NPV of 0.610) (Fig. 4). In multivariable regression analysis that included age, gender, A–elasticity, B–elasticity, B/A strain ratio, and the tumor area, A–elasticity identified 89.5% of malignancies (p < 0.001). A–elasticity and B–elasticity were the only significant independent factors influencing the identification of tumors (r^2 = 0.927, p < 0.001). It is worth noticing that the B/A strain ratio was eliminated during the regression analysis, as not significantly connected.

4   Discussion

This study demonstrates a significant connection between the parameters of A–elasticity, B–elasticity, B/A strain ratio, and malignant masses. The largest correlation coefficient was observed in the relationship between A–elasticity and the occurrence of malignancies. These data suggest an advantage of using only the A–elasticity value in differentiating malignancies from benign

Table 2 Elastography indices. Group 1 – malignant tumor (pancreatic adenocarcinoma) and Group 2 – benign tumor (chronic pancreatitis) patients

Index	All (n = 123)	Group 1 (n = 78)	Group 2 (n = 45)	p-value for Group 1 vs. Group 2
A–elasticity (%)	0.111 ± 0.121	0.024 ± 0.012	0.260 ± 0.063	<0.001
B–elasticity (%)	1.083 ± 0.549	0.862 ± 0.346	1.465 ± 0.624	<0.001
SR	28.051 ± 22.361	41.052 ± 17.973	5.517 ± 1.455	<0.001
Tumor area (cm^2)	9.50 ± 5.81	10.18 ± 5.80	8.33 ± 5.69	0.043
Mean maximal tumor diameter (cm)	3.77 ± 1.33	3.91 ± 1.15	3.53 ± 1.05	0.050

A–elasticity, elasticity of tumor area; B–elasticity, elasticity of reference zone area; SR, B/A strain ratio

Table 3 Associations of significance between the indices investigated (Spearman's correlation coefficient; $p < 0.05$)

	A–elasticity	B–elasticity	B/A strain ratio
Age	−0.309	ns	0.404
Gender (F)	−0.180	ns	0.290
Tumor area	−0.18	ns	0.245
Malignancy	0.841	0.487	−0.834
A–elasticity	–	0.713	−0.829
B–elasticity	0.713	–	0.241
SR	−0.829	−0.241	–

A–elasticity, elasticity of tumor area; B–elasticity, elasticity of reference zone area; SR, B/A strain ratio

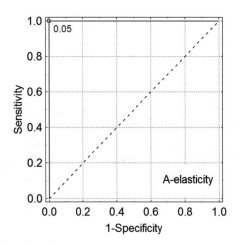

Fig. 1 Receiver operating characteristic (ROC) curve for A–elasticity

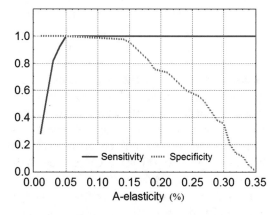

Fig. 2 A–elasticity: sensitivity and specificity for identifying malignant pancreatic tumors

tumors. The ROC analysis showed a similar discriminative ability of designated threshold values for both A–elasticity and the B/A ratio (sensitivity of 100% and specificity of 100%), indicating at least the same diagnostic usefulness of both. However, in regression analysis, A–elasticity was the main independent predictor of tumor malignancy, while the B/A strain ratio was not considered significant. These results suggest the possibility of using only the A–elasticity value to identify malignant pancreatic lesions, with no loss of sensitivity and specificity, which would allow to shorten the endoscopic diagnosis. Solid pancreatic lesions are most frequently detected incidentally by ultrasound of the abdomen. The

Fig. 3 B–elasticity/A–elasticity (B/A) strain ratio: sensitivity and specificity for identifying malignant pancreatic tumors

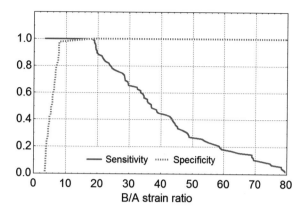

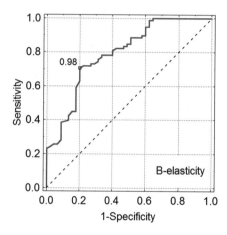

Fig. 4 Receiver operating characteristic (ROC) curve for B–elasticity

possibility of an adequate differentiation between malignant and benign pancreatic tumors based on Doppler perfusion parameters has been reported (Dyrla et al. 2016).

Currently, the standard procedure is the diagnosis using CT, magnetic resonance imaging, and endosonography (Sharma et al. 2011). Having in mind that in most cases of pancreatic cancer, treatment is still ineffective, and acute inflammation is often a revelator of pancreatic cancer, the methods enabling the accurate detection and differentiation of solid pancreatic lesions are necessary (Munigala et al. 2014; Ryan et al. 2014). Elastography is a method that enables a real-time imaging of tissue stiffness, and the test result is represented by a color image (red–yellow–green–blue) – an elastogram. The method is based on the assumption that elasticity of diseased tissue differs from normal with respect to tissue hardness. The tissue strain is evaluated based on the difference in the ultrasonic waves received (Giovannini 2009). In elastography, the breast is one of the first and most frequently assessed organs (Itoh et al. 2006). A rapid development of ultrasonography and endoscopy has enabled the introduction of elastography in endosonography. The assessment of mass elasticity is performed in real time. The result is presented in the form of a color image applied to a typical B-mode image (Săftoiu and Vilman 2006).

Endosonography is the best method for detecting solid pancreatic masses and connected with a fine needle aspiration (EUS–FNA) is characterized by a high sensitivity and specificity in differentiating pancreatic tumors (Iglesias–García and Domínguez–Muñoz 2007; Varadarajulu et al. 2005; Harewood and Wiersema 2002; Chang et al. 1997). In a multi-center study, Giovannini et al. (2009) have used elastography to assess focal pancreatic lesions detected during EUS in 121 patients. Seventy-two patients were diagnosed with adenocarcinoma, 16 with neuroendocrine tumor, 3 with metastases from other organs, and 30 with chronic pancreatitis, with 80.6% specificity and 92.3%. In another study, Săftoiu et al. (2008) have used elastography for the assessment of focal pancreatic lesions detected during EUS in 68 patients, who were then subjected to a 6-month-long follow-up observation. Thirty-two patients were diagnosed with pancreatic adeno-carcinoma, 11 with chronic pancreatitis, and 3 with pancreatic neuroendocrine tumors, and in 22 the organ was unchanged. That study employed a special computer software to objec-tify the assessment of the pathologies observed. The software enabled the digital development of a color histogram within the pathology detected and the determination of tissue hardness, yielding 91.4% sensitivity and 88.9% sensitivity of the test. In a study of Iglesias–Garcia et al. (2009), sensitivity of elastography was rated as very high compared to the reference methods, amounting to 100%, while specificity was 85.5%. Different results have been obtained by Hirche et al. (2008), who draw attention to a number of difficulties connected with elastography. The size of evaluated tumor appeared a significant limitation of the method. In case of a tumor larger than 3.5 cm, the method failed to embrace the whole tumor and to select a sufficiently large area of healthy tissue surrounding the lesion, which is used as a reference. Accordingly, sensitivity of the B/A strain ratio was rated at 41% and speci-ficity at 53%.

One of the most interesting reports on the use of elastography in the differential diagnosis of solid pancreatic tumors is a study that employed software enabling the quantitative assessment of tissue elasticity (Iglesias–Garcia et al. 2010). In that study, the assessment of tumor elasticity and B/A strain ratio was performed in each of the 86 patients. Sensitivity of quantitative elastography for the diagnosis of malignancy, using a B/A strain ratio of 6.04 as a cut-off value, was 100%, and specificity was 93%. Fur-ther, the authors also report that a cut-off value of 0.05 for the tumor mass elasticity (A–elasticity) differentiates malignant from benign lesions with sensitivity of 100% and specificity of 88.9%. These data are partly consistent with the present study as the two methods of evaluation: A–elas-ticity and B/A strain ratio yielded comparable results, with an advantage of A–elasticity over B/A strain ratio. In our study, the cut-off values of strain parameters were not arbitrarily chosen, but were set on the basis of the receiver operating curve analysis. We demonstrate that A–elasticity

was a somehow better predictor of pancreatic malignancies than the B/A strain ratio. Yet the reference of B–elasticity to A–elasticity value tended to be a more independent parameter. On the other hand, correlation of B–elasticity with malignancies suggests that the reference area is not objective and calls into question the assumptions of the method. Moreover, according to our observations and reports of other authors, the size of tumor exceeding 3.5 cm in diameter reduces the possibility of an adequate evaluation of B–elasticity, which further limits the useful-ness of the assessment of the B/A strain ratio, which is in contradistinction to the exclusive evaluation of A–elasticity (Hirche et al. 2008). In the Kongkam et al. (2015) study, negative results of both EUS–FNA and B/A strain ratio appear more reliable to exclude malignant solid pancreatic tumor, and the sensitivity of EUS elastography by B/A strain ratio is not superior to EUS–FNA.

Currently, there are two types of pancreatic elastography based on different principles, which are the strain elastography and the shear wave elastography. However, the latter cannot be applied in EUS to date (Kawada and Tanaka 2016). In the present study, A–elasticity between 0.05% and 0.14% identified malignancies with the highest sensitivity and specificity. Similar results have been obtained by Iglesias–Garcia et al. (2010) who have used the equipment of the same company. Unlike A–elasticity, the cut-off value for the B/A strain ratio of the highest sensitivity and specificity we found in the present study varied greatly compared to that indicated by those authors (range 7.87–18.23 vs. 6.04, respec-tively), which further supports the exclusive eval-uation of A–elasticity. Moreover, variability of the B/A strain ratio cut-off values suggests the necessity for defining a clinical center-specific cut-off value, contrary to A strain that seems to be center independent. This notion is supported by a significant relation of B strain to A strain shown for the first time in the presented study and by the difficulties with obtaining reliable values of B strain. On the basis of the literature and the present findings, we propose an isolated assess-ment of A strain in the identification of malignant pancreatic tumors. Nonetheless, this is a single-center study in which the measurements were performed by a single operator, which has a limiting effect on the results. A–elasticity and B/A strain ratio cut-off values obtained in this study, differentiating malignant from benign tumors, may vary in comparison to studies performed with other devices and by different operators. Therefore, we suggest to verify differential A–elasticity cut-off and B/A strain ratio values in larger groups of patients or in various clinical centers.

In conclusion, the present study suggests the use of the A–elasticity value obtained during endosonography examination in differentiating malignant from benign pancreatic tumors. The single parameter assessment of tumor mass elasticity is easier to obtain in elastography and appears sufficient to identify malignant tumors, without any loss of sensitivity and specificity of the test, giving an additional advantage of a faster endoscopic diagnosis.

Acknowledgments Funded by grant no. 245 from the Military Institute of Medicine in Warsaw, Poland.

Conflicts of Interest The authors declare no conflicts of interests in relation to this article.

Ethical Approval All procedures performed in studies involving human participants were in accordance with the ethical standards of the institutional and/or national research committee and with the 1964 Helsinki Declaration and its later amendments or comparable ethical standards. This study was approved by the Ethics Committee of the Military Institute of Medicine in Warsaw, Poland.

Informed Consent Written informed consent was obtained from all individual participants included in the study.

References

Aslam R, Yee J (2006) MDCT of pancreatic masses. Appl Radiol 35:4

Cascinu S, Falconi M, Valentini V, Jelic S, ESMO Guidelines Working Group (2010) Pancreatic cancer: ESMO Clinical Practice Guidelines for diagnosis, treatment and follow–up. Ann Oncol 21:55–58

Catalano MF, Sahai A, Levy M, Romagnuolo J, Wiersema M, Brugge W, Freeman M, Yamao K, Canto M, Hernandez LV (2009) EUS–based criteria for the diagnosis of chronic pancreatitis: the Rosemont classification. Gastrointest Endosc 69:1251–1261

Chang KJ, Nguyen P, Erickson RA, Durbin TE, Katz KD (1997) The clinical utility of endoscopic ultrasound–guided fine–needle aspiration in the diagnosis and staging of pancreatic carcinoma. Gastrointest Endosc 45:387–393

Davis SL, Brooke RJ, Kamaya A (2009) Islet–cell tumors of the pancreas: spectrum of MDCT findings. A pictorial essay. Appl Radiol 38:11

Dyrla P, Lubas A, Gil J, Niemczyk S (2016) Doppler tissue perfusion parameters in recognizing pancreatic malignant tumors. J Gastroenterol Hepatol 31:691–695

Giovannini M (2009) Contrast–enhanced endoscopic ultrasound and elastosonoendoscopy. Best Pract Res Clin Gastroenterol 23:767–779

Giovannini M, Botelberge T, Bories E, Pesenti C, Caillol F, Esterni B, Monges G, Arcidiacono P, Deprez P, Yeung R, Schimdt W, Schrader H, Szymanski C, Dietrich C, Eisendrath P, Van Laethem JL, Devière J, Vilmann P, Săftoiu A (2009) Endoscopic ultrasound elastography for evaluation of lymph nodes and pancreatic masses: a multicenter study. World J Gastroenterol 15:1587–1593

Harewood GC, Wiersema MJ (2002) Endosonography–guided fine needle aspiration biopsy in the evaluation of pancreatic masses. Am J Gastroenterol 97:1386–1391

Hirche TO, Ignee A, Barreiros AP, Schreiber–Dietrich D, Jungblut S, Ott M, Hirche H, Dietrich CF (2008) Indications and limitations of endoscopic ultrasound elastography for evaluation of focal pancreatic lesions. Endoscopy 40:910–917

Iglesias–García J, Domínguez–Muñoz JE (2007) Endoscopic ultrasound-guided biopsy for the evaluation of pancreatic tumors. Gastroenterol Hepatol 30:597–601

Iglesias–Garcia J, Larino–Noia J, Abdulkader I, Forteza J, Dominguez–Munoz JE (2009) EUS elastography for the characterization of solid pancreatic masses. Gastrointest Endosc 70:1101–1108

Iglesias–Garcia J, Larino–Noia J, Abdulkader I, Forteza J, Dominguez–Munoz JE (2010) Quantitative endoscopic ultrasound elastography: an accurate method for the differentiation of solid pancreatic masses. Gastroenterology 139:1172–1180

Itoh A, Ueno E, Tohno E, Kamma H, Takahashi H, Shilna T, Yamakawa M, Matsumura T (2006) Breast disease: clinical application of US elastography for diagnosis. Radiology 239:341–350

Kawada N, Tanaka S (2016) Elastography for the pancreas: current status and future perspective. World J Gastroenterol 22:3712–3724

Kongkam P, Lakananurak N, Navicharern P, Chantarojanasiri T, Aye K, Ridtitid W, Kritisin K, Angsuwatcharakon P, Aniwan S, Pittayanon R, Sampatanukul P, Treeprasertsuk S, Kullavanijaya P,

Rerknimitr R (2015) Combination of EUS–FNA and elastography (strain ratio) to exclude malignant solid pancreatic lesions: a prospective single–blinded study. J Gastroenterol Hepatol 30:1683–1689

Merkle EM, Bender GN, Brambs HJ (2000) Imaging findings in pancreatic lymphoma: differential aspects. Am J Roentgenol 174:671–675

Munigala S, Kanwal F, Xian H, Scherrer JF, Agarwal B (2014) Increased risk of pancreatic adenocarcinoma after acute pancreatitis. Clin Gastroenterol Hepatol 12:1143–1150

Perez–Johnston R, Sainani NI, Sahani DV (2012) Imaging of chronic pancreatitis (including groove and autoimmune pancreatitis). Radiol Clin N Am 50:447–466

Ryan DP, Hong TS, Bardeesy N (2014) Pancreatic adenocarcinoma. N Engl J Med 371:1039–1049

Săftoiu A, Vilman P (2006) Endoscopic ultrasound elastography – a new imaging technique for the visualization of tissue elasticity distribution. J Gastrointestin Liver Dis 15:161–165

Săftoiu A, Vilmann P, Gorunescu F, Gheonea DI, Gorunescu M, Ciurea T, Popescu GL, Iordache A, Hassan H, Iordache S (2008) Neural network analysis of dynamic sequences of EUS elastography used for the differential diagnosis of chronic pancreatitis and pancreatic cancer. Gastrointest Endosc 68:1086–1094

Scatarige JC, Horton KM, Sheth S, Fishman EK (2001) Pancreatic parenchymal metastases: observations on helical CT. Am J Roentgenol 176:695–699

Seufferlein T, Bachet JB, Van Cutsem E, Rougier P, on behalf of the ESMO Guidelines Working Group (2012) Pancreatic adenocarcinoma: ESMO–ESDO Clinical Practice Guidelines for diagnosis, treatment and follow–up. Ann Oncol 23:33–40

Sharma C, Eltawil KM, Renfrew PD, Walsh MJ, Molinari M (2011) Advances in diagnosis, treatment and palliation of pancreatic carcinoma: 1990–2010. World J Gastroenterol 17:867–897

Shimosegawa T, Chari ST, Frulloni L, Kamisawa T, Kawa S, Mino–Kenudson M, Kim MH, Klöppel G, Lerch MM, Löhr M, Notohara K, Okazaki K, Schneider A, Zhang L, International Association of Pancreatology (2011) International consensus diagnostic criteria for autoimmune pancreatitis: guidelines of the International Association of Pancreatology. Pancreas 40:352–358

Theoharis C (2008) Mast cells and pancreatic cancer. N Engl J Med 358:1860–1861

Varadarajulu S, Tamhane A, Eloubeidi MA (2005) Yield of EUS–guided FNA of pancreatic masses in the presence or the absence of chronic pancreatitis. Gastrointest Endosc 62:728–736

Wiersema MJ, Hawes RH, Lehman GA, Kochman ML, Sherman S, Kopecky KK (1993) Prospective evaluation of endoscopic ultrasonography and endoscopic retrograde cholangiopancreatography in patients with chronic abdominal pain of suspected pancreatic origin. Endoscopy 25:555–564

Adv Exp Med Biol - Clinical and Experimental Biomedicine (2019) 4: 49–54
https://doi.org/10.1007/5584_2018_287
© Springer Nature Switzerland AG 2018
Published online: 27 October 2018

Factors Affecting Health-Related Quality of Life in Liver Transplant Patients

Anna Jagielska, Olga Tronina, Krzysztof Jankowski,
Aleksandra Kozłowska, Katarzyna Okręglicka, Paweł Jagielski,
Magdalena Durlik, Piotr Pruszczyk, and
Aneta Nitsch–Osuch

Abstract

This study seeks to evaluate the metabolic parameters such as body mass index (BMI), percentage of total body fat percentage (%BF), blood glucose, homeostatic index for quantification of insulin resistance and beta-cell function (HOMA-IR), sleep efficiency, and physical activity in liver transplant patients. The study group consisted of 24 male and 18 female patients, which enabled the inter-gender comparison. We found that a majority of patients had exceeded the norms for BMI and %BF. The excessive weight was distinctly accentuated in male patients. Only 40.5% of patients have a correct BMI and 21.4% of patients have a correct %BF. The indices of glucose metabolism were increased, pointing to enhanced insulin resistance. Resting energy expenditure and metabolic equivalent of task were characteristic of sedentary lifestyle, and they were lower in female patients. Almost 65% of patients had sleep efficiency below the desired 85% cut-off level. Further, sleep efficiency was decreasing with increasing BMI, %BF, and blood glucose level. In conclusion, liver transplant patients are characterized by excessive body mass and less physical activity and have a shortened sleep duration, all of which may lead to a worse glucose metabolism and increased disease risk and may also have an impact on quality of life.

A. Jagielska, A. Kozłowska, K. Okręglicka, and
A. Nitsch–Osuch
Department of Social Medicine and Public Health,
Warsaw Medical University, Warsaw, Poland

O. Tronina (✉) and M. Durlik
Department of Transplant Medicine and Nephrology,
Warsaw Medical University, Warsaw, Poland
e-mail: msizp@wum.edu.pl

K. Jankowski and P. Pruszczyk
Department of Internal Medicine and Cardiology, Warsaw
Medical University, Warsaw, Poland

P. Jagielski
Human Nutrition Department, Faculty of Health Science,
Medical College of the Jagiellonian University, Cracow,
Poland

Keywords

Diabetes · Glycemia · Quality of life · Sleep efficiency · Metabolic status · Liver transplant

1 Introduction

Liver failure is a direct threat to life and results in liver transplantation in some qualified hepatologic patients, which provides improvement in patient's quality of life. Liver transplant patients become free of dietary restrictions present prior to surgery, and they often desire to regain the lost body weight, which might result in excessive weight gain. As compared to the general population, such patients have a higher prevalence of coronary artery diseases by 9–25%, hypertension by 60–70%, metabolic syndrome by 44–58%, type 2 diabetes by 30–40%, and dyslipidemia by 45–69% (Singh and Watt 2012).

Sleeping disorders affect 3–17% adults, and if untreated, they increase the risk of noncommunicable diseases (Shrivastava et al. 2014). Recommended sleep efficiency for adults is ≥85%/night (Landry et al. 2015). Sleep loss is associated with many metabolic consequences, such as obesity, impaired glucose tolerance or diabetes, heart attacks, strokes, adverse effects on mood and behavior (excess mental distress, depressive symptoms, anxiety, and alcohol use), and increased age-specific mortality (Poggiogalle et al. 2018; Colten and Altevogt 2006). Obesity due to sleep loss is a sequel of lower release of leptin by adipocytes, a peptide that restrains appetite, and increased release of ghrelin, a peptide that stimulates appetite. Disruption of the circadian system impairs the central clock based on melatonin, cortisol, and core body temperature. That increases the risk of noncommunicable diseases, since circadian misalignment elevates glucose, insulin, and triglyceride levels, which is accompanied by a lower energy expenditure and impaired peripheral insulin sensitivity (Poggiogalle et al. 2018).

The aim of the study was to evaluate the metabolic parameters such as body mass index (BMI), percentage of total body fat (%BF), blood glucose, glycated hemoglobin A1 (HbA1c), homeostatic model for quantification of insulin resistance and beta-cell function (HOMA-IR), sleep efficiency, and physical activity in liver transplant patients.

2   Methods

The study group consisted of 42 post-transplant patients (18 women and 24 men), investigated from September 2015 to May 2016 at the Department of Transplant Medicine and Nephrology, Warsaw Medical University in Warsaw, Poland.

Liver transplantation had been performed in these patients between 2003 and 2014.

Anthropometric measurements consisted of weight, height, %BF, and resting energy expenditure (REE) (BioScan 920-2S Multi-frequency Analyzer, Maltron International Ltd., Rayleigh, Essex, UK). Body mass index (BMI) was assessed in accordance with the WHO guidelines, where normal BMI range is 18.5–24.9 kg/m^2 (WHO 2003). %BF >25% in men and > 30% in women indicated metabolic obesity (Zeng et al. 2016; Suliga 2012). Biochemical blood indices were obtained from the medical history files, as the patients underwent routine examinations during regular control visits after liver transplantation at the Department of Transplant Medicine and Nephrology, Warsaw Medical University in Warsaw.

The SenseWear Pro Armband^{MT} (BodyMedia, Pittsburgh, PA) was used to continuously record energy expenditure, a surrogate of physical activity, assessed as the metabolic equivalent of task (1 MET = 1 kcal/kg/h or 3.5 ml O_2/kg/min), body motion, and sleep efficiency. The device is a dual axis accelerometer that noninvasively senses both static and dynamic forces of motion and records rest/activity cycles. The sensor was worn on the triceps muscle of an arm by patients for 5–7 days. Sensitivity of the recordings enabled the differentiation among the conditions of sleep, wakefulness, and lying down but awake (Liden et al. 2002). Total time in bed at night was defined as the length of time from lights off to lights on. Sleep efficiency was calculated as the percentage of time asleep of the total time staying in bed. The evaluation of sleep efficiency using the SenseWear Pro Armband^{MT} has been found to closely match that obtained from full night polysomnography, the gold standard for the estimation of sleep time and its disorders (Sharif and BaHammam 2013).

Data were presented as mean ± SD, when normally distributed, or median and min–max range, when non-normally distributed, and as percentages. A t-test, chi-square, and Mann–Whitney U test were used for statistical comparisons. Spearman's correlation coefficient was used to evaluate associations between parameters. A p-value <0.05 defined the statistically significant differences. Statistical elaboration was performed with a commercial SPSS statistical package (IBM Corp, Armonk, NY).

3 Results

The group of liver transplant patients consisted of 42 subjects (18 women and 24 men) of the mean age 48.6 ± 11.1 (range 27–68 years), body weight 79.2 ± 16.3 kg, BMI 27.7 ± 5.0 kg/m², and %BF 29.6 ± 10.5%. The median resting energy expenditure was 1,612.5 kcal (range 1,227–2,433 kcal/24 h), whereas physical activity amounted to 1.5 METs (0.9–2.0) kcal/kg/h. Significant differences between genders were observed for weight, height, resting energy expenditure, and glycemia (Table 1).

Normal BMI was present in 40.5% of liver transplant patients (Fig. 1) and normal total body fat in just 21.4% of patients (Fig. 2). Almost 65% of patients had sleep efficiency below the recommended 85% *per* night. The mean sleep efficiency in patients with normal BMI and normal %BF approximated 85%. Sleep efficiency in overweight and obese patients and in those with %BF below or above the norm was decreased, but differences were insignificant when compared to normal body weight and normal %BF (Figs. 3 and 4). No gender differences were noticeable in BMI, percentage of total body fat. Gender-related differences were, however, noticeable in sleep efficiency in liver transplant patients. Sleep efficiency was significantly inversely associated with BMI, %BF, and blood glucose level in men, but not in women, and with HOMA-IR in women only after liver transplant (Table 2).

4 Discussion

In this study we found that the majority of liver transplant patients of 59.5% had excessive body weight, one-third was overweight and one-fourth was obese, having the mean weight of 79.2 ± 16.3 kg and BMI of 27.7 ± 5.0 kg/m². That corresponds closely to the 58.6% of excessive body weight in the Polish general

Table 1 Anthropometric, physical activity, sleep efficiency, and biochemical data of liver transplant patients

	Men		Women			Both genders	
	Mean ± SD or median (min-max)	n	Mean ± SD or median (min-max)	n	p <	Mean ± SD or median (min-max)	n
Age (year)	48.4 ± 12.0	24	49.0 ± 10.2	18	ns	48.6 ± 11.1	42
Weight (kg)	85.5 ± 15.9	24	70.8 ± 13.0	18	0.003	79.2 ± 16.3	42
Height (cm)	176.6 ± 10.6	24	159.3 ± 7.3	18	0.0001	169.2 ± 12.7	42
BMI (kg/m²)	27.5 ± 5.1	24	28.1 ± 5.1	18	ns	27.7 ± 5.0	42
%BF	24.1 ± 5.1	24	36.9 ± 8.8	18	ns	29.6 ± 10.5	42
REE (kcal/24 h)	1,892 (1,499–2,433)	24	1,407 (1,227–1,665)	18	0.0001	1,613 (1,227–2,433)	42
MET (kcal/kg/h)	1.6 (0.9–2.0)	24	1.4 (1.0–1.9)	18	ns	1.5 (0.9–2.0)	42
SE (%)	80.0 ± 8.0	24	84.0 ± 7.0	18	ns	81.0 ± 8.0	42
BG (mg/dL)	100 (74–187)	24	88.5 (60–103)	18	0.032	92 (60–187)	42
HbA1c (%)	5.3 (4.1–6.5)	21	5.0 (4.4–5.9)	16	ns	5.2 (4.1–6.5)	37
Insulin (μIU/mL)	7.6 (2.7–22.0)	20	6.9 (3.5–17.9)	17	ns	7.6 (2.7–22.0)	37
HOMA-IR	1.05 (0.38–3.20)	20	0.90 (0.45–2.35)	17	ns	1.02 (0.40–3.20)	37

BMI body mass index, *%BF* percentage of total body fat, *REE* resting energy expenditure, *MET* metabolic equivalent of task, *SE* sleep efficiency, *BG* blood glucose, *HbA1c* glycated hemoglobin A1, *HOMA–IR* homeostatic model for quantification of insulin resistance, *ns* nonsignificant; p-value denotes inter-gender difference

%

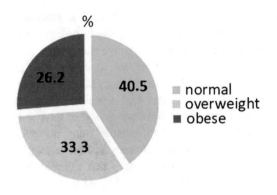

Fig. 1 Body mass index (BMI) in liver transplant patients

%

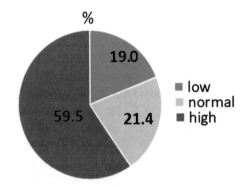

Fig. 2 Percentage of total body fat (%BF) in liver transplant patients

population (WHO 2013) and the findings of body weight and BMI in the American population of liver transplant patients aged below 65 (85 kg and 28 kg/m^2, respectively) reported by Wang et al. (2015). Excessive body weight was reported in 84% of liver patients by Bulzacka (2008), and overweight and obesity in 28% and 50% patients, respectively, by da Silva Alves et al. (2014). We found that 40.5% of our liver patients had BMI within the WHO norm. On the other hand, almost 60% of our liver transplant patients had an excessive percentage of total body fat, with the mean of 29.6 ± 10.5%, while the Dutch and American studies demonstrate a greater percentage of body fat, amounting to 26.9 ± 6.4% in patients aged 48.0 ± 11.8 and 32.5 ± 9.6% in those aged 50.1 ± 13.1 (Ribeiro et al. 2014; Berbke et al. 2007).

Physical activity is essential for the proper functioning of a human body and for the prevention of noncommunicable diseases. Patients with regular physical activity show a greater vitality and experience less restrictions in their daily-life functioning (Painter et al. 2001). Reduced physical activity has been observed in studies examining the quality of life of liver transplant patients (Kotarska et al. 2015a, b; Ribeiro et al. 2014), which has been shown to increase the risk of noncommunicable diseases and sleeping disorders (Poggiogalle et al. 2018; WHO 2003). In the present study, resting energy expenditure was lower in the liver transplant patients. Physical activity amounted, on average, to 1.5 METs (range 0.9–2.0 METs), which is characteristic of the sedentary lifestyle. Anastacio et al. (2011) have reported a mean value of MET for liver transplant patients of 1.35 ±0.17, which is comparable to the present findings.

Chronobiology is at play in the development of noncommunicable diseases such as diabetes, obesity, or hypertension. Circadian rhythms are essential regulators of glycemia, insulin resistance, glucose tolerance, lipid profile, energy expenditure, and appetite (Poggiogalle et al. 2018). Sleep quality is essential for the feeling of well-being and for maintaining the pro-health behaviors. Sleeping disorders affect 3–17% adults and, if untreated, they increase the risk of noncommunicable and mental disorders (Colten and Altevogt 2006). Recommended sleep efficiency for adults is ≥85% of nighttime (Markwald et al. 2016). The prevalence of a short sleep duration (<5 h daily) in the US adult population has increased from 1.7% in 1977 to 2.4% in 2009, which is associated with impaired regulation of blood glucose – odds ratio (OR) 2.15 (1.85–2.50), and increased risk of overweight – OR 1.30 (1.19–1.43) or obesity – OR 2.17 (1.97–2.40) (Girardin et al. 2014). A deleterious effect on glycemic control of circadian rhythm misalignment has also been noticed in other studies (Poggiogalle et al. 2018). In the present study, only 65% of liver transplant patients had sleep efficiency above 85%, and sleep efficiency was inversely associated with overweight/obesity and excess percentage of total body fat in men, but not in women. Gender-related differences in sleep efficiency

Fig. 3 Sleep efficiency by the body mass index (BMI). Data are mean ± SD; n = 42. The intergroup differences in sleep efficiency were insignificant

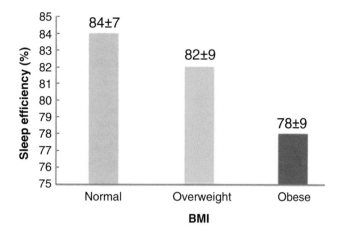

Fig. 4 Sleep efficiency by percentage of total body fat (%BF). Data are mean ± SD; n = 42. The intergroup differences in sleep efficiency were insignificant

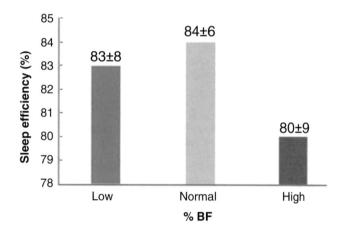

Table 2 Associations between the metabolic parameters investigated and sleep efficiency

	Sleep efficiency	
	r-value	p-value
BMI – men (n = 24)	−0.41	0.045
BMI – women (n = 18)	−0.097	0.703
%BF – men (n = 24)	−0.442	0.031
%BF – women (n = 18)	−0.027	0.915
Blood glucose – men (n = 24)	−0.437	0.033
Blood glucose – women (n = 18)	−0.117	0.636
HOMA-IR – men (n = 24)	−0.272	0.247
HOMA-IR – women (n = 18)	−0.620	0.008

BMI body mass index, *%BF* percentage of total body fat, *HOMA–IR* homeostatic model for quantification of insulin resistance, *r*-value Spearman's correlation coefficient; bold font indicates significant changes

were also noticeable concerning the glucose metabolism. Sleep efficiency associated inversely with increasing blood glucose and HOMA-IR index levels; the former was significantly accentuated in men and the latter, which translates into a greater insulin resistance, in women with liver transplants.

We conclude that liver transplant patients manifest disordered sleep duration and excessive body mass, which may worsen glucose metabolism, may increase disease risk, and may also have an impact on quality of life. The study points to the corrective steps to be taken in the management of patients after liver transplants toward healthy living and prevention of noncommunicable diseases, such patients are vulnerable to.

Acknowledgments Funded by the statutory budget of Warsaw Medical University in Poland.

Conflicts of Interest The authors declare no conflict of interests in relation to this article.

Ethical approval All procedures performed in studies involving human participants were in accordance with the ethical standards of the institutional and/or national research committee and with the 1964 Helsinki declaration and its later amendments or comparable ethical standards.

Informed consent Informed consent was obtained from all individual participants included in the study.

References

Anastacio LR, Ferreira LG, Ribeiro HS, Liboredo JC, Lima AS, Correia MI (2011) Metabolic syndrome after liver transplantation: prevalence and predictive factors. Nutrition 27:931–937

Berbke TJ, van den Berg–Emons RJG, Kazemier G, Metselaar HJ, Tilanus HW, Stam HJ (2007) Physical fitness, fatigue, and quality of life after liver transplantation. Eur J Appl Physiol 100:345–353

Bulzacka M (2008) Quality and organization of patient's care after organ transplantation. Nurs Top 16 (1, 2):54–59 Article in Polish

Colten HR, Altevogt BM (2006) Sleep disorders and sleep deprivation: an unmet public health problems. Institute of Medicine (US) Committee on Sleep Medicine and Research. The National Academies Press, Washington, DC. https://doi.org/10.17226/11617. Accessed on 22 July 2018

da Silva Alves V, Mendes RH, Pinto Kruel CD (2014) Nutritional status, lipid profile and HOMA–IR in post–liver transplant patients. Nutr Hosp 29(5):1154–1162

Girardin JL, Williams NJ, Sarpong D, Pandey A, Youngsted S, Zizi F, Ogedegbe G (2014) Assosiation between inadequate sleep and obesity in the US adult population: analysis of the national health survey (1977–2009). BMC Public Health 14:290

Kotarska K, Wunsch E, Raszeja–Wyszomirska J, Kempińska–Podhorodecka A, Wójcicki M, Milkiewicz P (2015a) Leisure time physical activity and health–related behaviors after liver transplantation: a prospective, single–centre study. Prz Gastroenterol 10(2):100–104

Kotarska K, Wunsch E, Raszeja–Wyszomirska J, Kempińska–Podhorodecka A, Wójcicki M, Milkiewicz P (2015b) Female sex but not original indication affects physical activity after liver

transplant: a prospective, single centre study. Exp Clin Transplant 13(3):243–246

Landry GJ, Best JR, Liu–Ambrose T (2015) Measuring sleep quality in older adults: a comparison using subjective and objective methods. Front Aging Neurosci 7:166

Liden CB, Wolowicz M, Stivoric J, Teller A, Vishnubhatla S, Pelletier R, Farringdon J (2002) Accuracy and reliability of the SenseWear™ armband as energy expenditure assessment device. http://www. bodymedia.com. Accessed on 23 July 2018

Markwald RR, Bessman SC, Reini SA, Drummond SPA (2016) Performance of a portable sleep monitoring device in individuals with high versus low sleep efficiency. J Clin Sleep Med 12(1):95–103

Painter P, Krasnoff J, Paul SM, Ascher NL (2001) Physical activity and health–related quality of life in liver transplant recipients. Liver Transpl 7:213–219

Poggiogalle E, Jamshed H, Peterson CM (2018) Circadian regulation of glucose, lipid and energy metabolism in humans. Metabolism 84:11–27

Ribeiro HS, Anastácio LR, Ferreira LG, Lima AS, Correia MI (2014) Energy expenditure and balance among long term liver recipients. Clin Nutr 33:1147–1152

Sharif MM, BaHammam AS (2013) Sleep estimation using BodyMedia's SenseWear™ armband in patients with obstructive sleep apnea. Ann Thorac Med 8 (1):53–57

Shrivastava D, Jung S, Saadat M, Sirohi R, Crewson K (2014) How to interpret the results of a sleep study. J Community Hosp Intern Med Perspect 4(5):24983

Singh S, Watt KD (2012) Long–term medical management of the liver transplant recipient: what the primary care physician needs to know. Mayo Clin Proc 87:779–790

Suliga E (2012) Abdominal obesity – methods of evaluation, causes of occurrence, health implications. Stud Med 27(3):65–71 Article in Polish

Wang CW, Covinsky KE, Feng S, Hayssen H, Segev DL, Lai JC (2015) Functional impairment in older liver transplantation candidates: from the functional assessment in liver transplantation study. Liver Transpl 21:1165–1470

WHO (2003) Diet, nutrition and the prevention of chronic diseases: report of a joint WHO/FAO expert consultation. Geneva. http://whqlibdoc.who.int/trs/WHO_TRS_916.pdf. Accessed on 6 May 2018

WHO (2013) Nutrition, physical activity and obesity in Poland. Available at: http://www.euro.who.int/__data/assets/pdf_file/0020/243317/Poland–WHO–Country–Profile.pdf?ua=1. Accessed on 30 Apr 2017

Zeng Q, Dong SY, Wang ML, Li JM, Ren CL, Gao CQ (2016) Obesity and novel cardiovascular markers in a population without diabetes and cardiovascular disease in China. Prev Med 91:62–69

Adv Exp Med Biol - Clinical and Experimental Biomedicine (2019) 4: 55–63
https://doi.org/10.1007/5584_2018_323
© Springer Nature Switzerland AG 2019
Published online: 29 January 2019

Phagocytosis and Autophagy in THP-1 Cells Exposed to Urban Dust: Possible Role of LC3-Associated Phagocytosis and Canonical Autophagy

A. Holownia, A. Niechoda, J. Lachowicz, E. Golabiewska, and U. Baranowska

Abstract

Exposure to ambient particulate matter (PM) increases mortality and morbidity due to respiratory and cardiovascular diseases. The aim of this study was to assess the effect of standardized urban dust (UD) on phagocytosis and autophagy in a monocyte-macrophage cell line (THP-1 cells). The cells were grown for 24 h in the medium supplemented with 200 $\mu g \cdot mL^{-1}$ coarse carbon black (CB) or UD. In some experiments glutathione (GSH) was depleted in THP-1 cells by buthionine sulfoximine. The cells were double stained with green latex beads (phagocytosis) and with red autophagy marker (LC3) and were evaluated in a flow cytometer. In naïve THP-1 cells, about 61% of them were classified as "negative", while 39% were classified as "double-positive". Both GSH depletion and UD treatment produced three distinct subpopulations of cells on bivariate scatterplots. A new subpopulation of cells (about 24% of the total number) appeared, with a lower autophagy and phagocytosis, but with a higher autophagy/phagocytosis ratio, when compared to highly positive cells. CB affected, to some extent, phagocytosis without a substantial effect on autophagy. In conclusion, the research on distinct pathways of immune cell activation may be relevant to the diagnostics and therapy of PM-induced pneumotoxicity, inflammation, and tumorigenesis.

Keywords

Autophagy · Carbon black · Phagocytosis · THP-1 cells · Urban dust

1 Introduction

Exposure to ambient particulate matter (PM) increases mortality and morbidity of respiratory and cardiovascular diseases (Anderson et al. 2012). Environmental pollutants damage respiratory epithelial cells, causing inflammation and tissue injury (Bai et al. 2016). The molecular mechanisms of PM-mediated toxicity and immunotoxicity are unclear, yet oxidative stress and DNA damage are definitely involved (de Oliveira Alves et al. 2011, 2014). Apart from direct cytotoxicity, the early cellular response to environmental pollutants involves

A. Holownia (✉), A. Niechoda, J. Lachowicz,
E. Golabiewska, and U. Baranowska
Department of Pharmacology, Medical University,
Bialystok, Poland
e-mail: adam.holownia@umb.edu.pl

the activation of biochemical pathways that aim to reestablish homeostasis. Recent data indicate that autophagy, which is a lysosome-involving catabolic, self-digestion process, is relevant to PM-induced stress (Peixoto et al. 2017). It has been shown that protein kinase mechanistic target of rapamycin (mTOR), an element of autophagy signaling pathways, suppresses the PM-induced necroptosis, activation of NFκ light-chain enhancer of activated B cells (NF-κB), and the inflammatory response in lung macrophages (Zhang et al. 2012). However, the regulation of PM-induced inflammatory response is still unclear. Classical autophagy pathways can be divided into macroautophagy, microautophagy, and chaperone-mediated autophagy (Chen and Klionsky 2011), but some elements of autophagy participate in noncanonical cell functions, especially in pathology (Kalai Selvi et al. 2017). The signaling elements of autophagy can also be targeted specifically to damaged mitochondria (mitophagy), altered proteins (aggrephagy), and internalized pathogens (xenophagy) (Singh et al. 2018). A popular marker of autophagy, the microtubule-associated protein 1A/1B-light chain (LC3), can also be recruited to phagosomal membranes during LC3-associated phagocytosis, an unconventional form of autophagy (Heckmann et al. 2017). Of note, classical autophagy and probably also immune response may be impeded during noncanonical autophagy, resulting in changes in inflammatory signaling (Matsuzawa-Ishimoto et al. 2018). The aim of this study was to define the effect of standardized urban dust (UD) on phagocytosis and autophagy in THP-1 cells, a human monocyte-macrophage cell line.

2 Methods

2.1 Cell Culture

The human monocyte-macrophage cell line, THP-1 cells (TIB202™), was grown in American Type Culture Collection (ATCC)-formulated Roswell Park Memorial Institute (RPMI) 1640 medium, supplemented with 2-mercaptoethanol to a final concentration of 0.05 mM and with 10% FBS. Cells were maintained in 37° C in an incubator in a humidified atmosphere containing 5% CO_2. For the experiments, the cells were plated out onto 6- or 12-well plates and grown in control or conditioned media for 24 h.

2.2 Cell Treatment

Carbon black (CB) and urban dust (UD)-conditioned media were prepared using commercial, standardized UD purchased from the National Institute of Standards and Technology (Gaithersburg, MD) or with a coarse CB (260 nm diameter, Huber 990; Haeffner and Co. Ltd., Chepstow, UK), which was used as a reference substance. According to the Certificate of Analysis of Standard Reference Material 1649b, the particle size of UD was within a range of 0.20–110 μm, with a mean particle size of about 10 μm. For experiments, particles were suspended in RPMI 1640 medium at a concentration of 10--200 μg·mL^{-1} and sonicated in an ultrasonic homogenizer (Bandelin Sonoplus; Berlin, Germany) for 30 s prior to use. CB- and UD-conditioned media were used within 5 min of preparation. In some experiments, THP-1 cells were pretreated overnight with 100 μM buthionine sulfoximine (BSO) to deplete their content of glutathione (GSH). Cell-free controls were included to each experiment in order to assess the interference of particles with each assay.

2.3 Phagocytosis Assay

Cayman's Phagocytosis Assay Kit (Ann Arbor, MI), which employs latex beads coated with fluorescein isothiocyanate (FITC)-labeled rabbit IgG, was used as a probe at a bead-to-cell ratio of 10:1 for the quantification of phagocytosis in THP-1 cells. After incubation, cells were washed twice with assay buffer, and signals from the internalized florescent beads were detected on Fl1 channel of a flow cytometer (FACSCanto II, BD Biosciences Systems, San Jose, CA).

2.4   Autophagy Determination

The autophagy marker LC3 protein, which is recruited to the double membrane of an autophagosome, was used to quantify autophagy. LC3 probes were specific monoclonal LC3A/B rabbit antibodies, conjugated to Alexa Fluor 647 (Cell Signaling Technology, Danvers, MA). To quantify LC3 protein, THP-1 cells were fixed in 4% formaldehyde, permeabilized with 90% methanol, immunostained with specific antibodies for 1 h, washed twice with the incubation buffer, and analyzed in a flow cytometer.

2.5   Phagocytosis/Autophagy Assay

THP-1 cells were double stained using a typical protocol for phagocytosis. Cells were incubated for 2 h with FITC-labeled latex beads followed by 1 h incubation with anti-LC3 Alexa Fluor 647 antibody, using 4% formaldehyde and 90% methanol to permeabilize the cell membrane. The unstained THP-1 cells and single (green or red)-stained cells served as controls to calibrate the flow cytometer detectors and compensation. Samples were analyzed with a FACSCanto II flow cytometer (BD Biosciences Systems, San Jose, CA) with a standard filter setup. The green fluorescence from FITC was detected with a 530 nm, 30 nm bandwidth band-pass filter, in FL1 channel, and red fluorescence of the Alexa Fluor 647 was collected in FL3 channel (>600 nm long-pass filter). All analyses were performed at a low rate settings with <1000 events/s. The experimental data were plotted as bivariate cytograms, and then scatterplots and fluorescence histograms were analyzed using the Flowing Software v2.5 (Cell Imaging Core, Turku Center for Biotechnology, Turku, Finland).

2.6   Statistical Analysis

Results were expressed as means of 4–6 assays ±SD. Inter-group differences were evaluated with one-way or two-way ANOVA followed by Bonferroni post hoc tests for selected pairs of data. A p-value <0.05 defined statistically significant differences. Statistical analysis was performed with a commercial package of Statistica 6.0 software (StatSoft, Cracow, Poland).

3   Results

Preliminarily, time- and concentration-dependent experiments were done to establish the experimental conditions for double staining of cells for phagocytosis and autophagy. Firstly, different numbers of THP-1 cells were incubated for 1–6 h with FITC-labeled latex beads to choose a convenient incubation time for phagocytosis. With 50,000 cells/sample, 2 μL of latex beads bound to FITC and 2 h of incubation about 40% of THP-1 cells were positively stained with FITC, irrespective of the cell treatment. Thus, 2-h time was selected as the incubation time with a phagocytosis marker in the double-staining protocol. Moreover, 200 $\mu g \cdot mL^{-1}$ concentrations of CB and UD and 24-h exposure time were chosen due to a limited UD toxicity in THP-1 cells. Neither CB nor UD by themselves significantly altered red or green fluorescence signals. Changes in the cell number (% of cells) and the green/red fluorescence ratios (phagocytosis/autophagy) are shown in Table 1, while typical bivariate cytograms of representative cell treatments are shown in Fig. 1.

Control naïve THP-1 cells are shown in Fig. 1-panel a, cells grown for 24 h with 200 $\mu g \cdot mL^{-1}$ CB in panel b, cells co-incubated with 200 $\mu g \cdot mL^{-1}$ UD in panel c, and cells pretreated with 100 μM BSO and then treated with CB or UD are in panels d, e, and f, respectively. There are two subpopulations of control THP-1 cells visible (Fig. 1-panel a): "negative" for both phagocytosis and autophagy in quadrant 3 and "positive" for both FITC (phagocytosis) and Alexa Fluor (autophagy) in quadrant 2. In our experimental conditions, about 61% of control cells were "negative" (panel a, quadrant 3), and 39% of cells were "positive" (quadrant 2; panel a). The corresponding relative green and red fluorescences measured in "positive" control

Table 1 The effects of carbon black (CB) and urban dust (UD) on phagocytosis and autophagy (LC3A/B expression) in human monocyte-macrophage cell line – THP-1 cells. Cells were grown for 24 h in the culture medium supplemented with 200 μg·mL^{-1} CB or UD. In some experiment, cells were pretreated for 24 h with 100 μg·mL^{-1} buthionine sulfoximine (BSO) to deplete glutathione. Phagocytosis and autophagy were quantified in flow cytometry in double-(green/red)-stained cells, and results were expressed as % of positive or negative cells. Median green/red fluorescence intensities were calculated and are shown as phagocytosis/autophagy ratios for different cell subpopulations

	Negative (%)	Low-positive		High-positive	
		Fraction (%)	Ratio	Fraction (%)	Ratio
Control	61 ± 8.8	4 ± 2.5		39 ± 5.2	0.40 ± 0.06
BSO	42 ± 7.4**	13 ± 4.1**	0.20 ± 0.04	45 ± 7.6	0.32 ± 0.05
CB	53 ± 7.1##	3 ± 2.6##		47 ± 5.3*	0.40 ± 0.06
UD	38 ± 5.6**	24 ± 5.3**##^^	0.20 ± 0.05	38 ± 4.9##	0.22 ± 0.03**^
BSO + CB	54 ± 6.9	3 ± 3.3##		46 ± 5.9*	0.41 ± 0.05
BSO + UD	16 ± 8.1**##^^	23 ± 4.2**##	0.11 ± 0.04#^	61 ± 8.2**##^^	0.75 ± 0.08**##^^

*p < 0.05; **p < 0.01 – for comparisons with the corresponding control cells
##p < 0.01 – for comparisons with the corresponding BSO-treated cells
^p < 0.05; ^^p < 0.01 – for comparisons with CB or UD, respectively

cells were expressed as 100 relative units, and the mean fluorescence ratios were calculated for each subpopulation of cells (Table 1). In the cells with depleted GSH content (cells pretreated overnight with BSO; Fig. 1-panel d), in the cells grown with UD (panel c), and in the cells with depleted GSH and then treated with UD (panel f), there appears another cell subpopulation in quadrant 4 (see also Table 1 – "low-positive" cells). These cells have a lower, but still positive, both phagocytic activity and autophagy when compared to the cells in quadrant 3, but they have a higher LC3/phagocytosis ratio, which points to a more intense autophagy than phagocytosis. The results were expressed as percentages of the total (positive and negative) cell numbers. The corresponding statistical data (frequency, standard deviations, and relative green/red fluorescence ratios) for each group are shown in Table 1.

Considering the population data, cell treatment with BSO and independently with UD decreased the "negative" (no phagocytosis and no autophagy) cell numbers by 31% (BSO; p < 0.01), 38% (UD; p < 0.01), and 74% (BSO + UD; p < 0.01) when compared to the 61% of control "negative" cells (panel a, quadrant 3). On the other hand, BSO increased the "low-positive" cell numbers by more than threefold (p < 0.01), while UD increased the "low-positive" cell numbers by about sixfold (p < 0.01), and when both compounds were applied together, the number of phagocytic/autophagic cells increased about sixfold for "low-positive" cell fraction and by about 56% (p < 0.01) for "high-positive" cells. Of note, active THP-1 cells grown with BSO and/or UD, but not with CB, had binary distribution. On the other hand, cell treatment with CB resulted in an insignificant decrease in the "negative" cell numbers and in a small significant increase in the "high-positive" cell numbers (20%; p < 0.05). Interestingly, in the presence of CB, the effect of BSO on the "low-positive" cells was no longer observed.

Considering quantitative changes in phagocytosis and LC3 expression in two subpopulations of "positive" cells, the ratio of green-to-red relative fluorescence intensities, i.e., phagocytosis/autophagy in the control highly positive cells, was 0.40. The ratio decreased by BSO pretreatment to 0.32 and to 0.22 (p < 0.01) in cells grown with UD, which may indicate the relative prevalence of autophagy over phagocytosis. Unexpectedly, in the cells treated with both compounds, this ratio increased to 0.75 (p < 0.01) in the highly "positive" cells. At the same time, the numbers of "low-positive" cells with lower fluorescence intensity and with phagocytosis/autophagy ratio of 0.11 increased to about 23% (p < 0.01) of the total cell number. Our data indicate that these new cells originated from the unstained "negative" cells, but not from "high-positive" THP-1 cells.

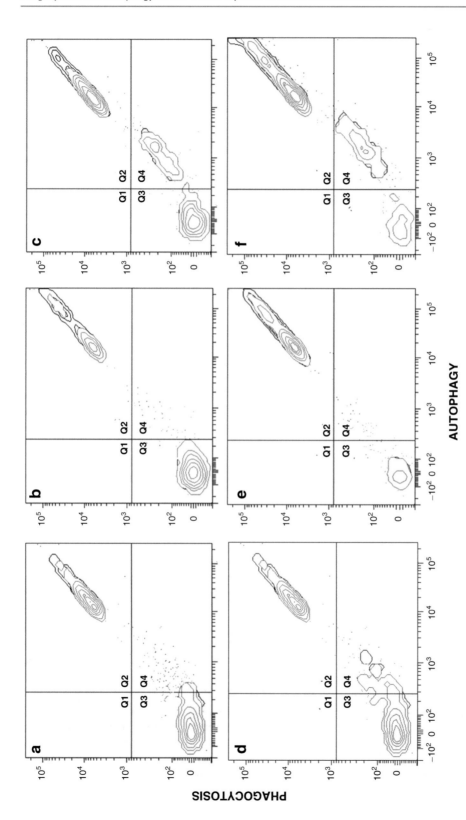

Fig. 1 The effects of carbon black (CB) and urban dust (UD) on phagocytosis and autophagy (LC3A/B expression) in human monocyte-macrophage cell line – THP-1 cells. Cells were grown for 24 h in the culture medium supplemented with 200 µg·mL^{-1} CB or UD. In some experiments, cells were pretreated for 24 h with 100 µg·mL^{-1} buthionine sulfoximine (BSO) to deplete glutathione. Phagocytosis and autophagy were quantified in flow cytometry in double-(green/red)-stained cells, and the results were expressed as % of positive or negative cells. (**a**) Control naïve THP-1 cells, (**b**) cells treated with CB, (**c**) cells treated with UD, (**d**) cells treated with BSO, (**e**) cells treated with BSO + CB, and (**f**) cells treated with BSO + UD

4   Discussion

Environmental pollutants damage respiratory cells, cause inflammation, and increase the incidence of respiratory and cardiovascular diseases. Due to heterogeneous, time- and place-dependent composition of PM, changeability of exposure, and highly variable clinical presentations such as acute and chronic toxicity, inflammation, autoimmunity, and tumorigenesis, the clinical and experimental problem is complex. Moreover, there is no valuable diagnostic and predictive biomarker that can be used in monitoring the risk of PM exposures and disease progression.

Mononuclear phagocytes are protecting cells of the respiratory tract. They have the ability to phagocyte foreign compounds and pathogens and play important roles in innate and adaptive immunity and inflammation (Baharom et al. 2017). The activation of phagocytes may be caused not only by viruses and bacteria (Dale et al. 2008) but also by different chemical stimuli, including PM (Alexis et al. 2006). THP-1 cells, a human monocyte-macrophage cell line, are widely used an in vitro model for investigating cell differentiation and changes in cell functions. THP-1 cells are redox dependent and are useful in the assessment of effects of PM, a powerful prooxidant (Lawal 2017). Oxidative stress can affect the physiology of immune cells. The intracellular redox balance is tightly regulated in macrophages, where reactive oxygen species are necessary to cell maturation, interleukin production, and phagocytosis (Li et al. 2018; Szabó-Taylor et al. 2017). THP-1 cells are especially fragile to redox imbalance. These cells require 2-mercaptoethanol-supplemented culture medium to protect their thiol-dependent homeostasis (Grodzki et al. 2013). Nonetheless, these cells are often used as a model to study human inflammatory responses (Nicholas and Sumbayev 2009; Dobrovolskaia et al. 2008).

The present study demonstrates that the response of THP-1 cells to CB and UD is different. The CB was able to affect, to an extent, phagocytosis of THP-1 cells without a substantial effect on autophagy, while UD disparately affected autophagy with only a limited effect on phagocytosis. Moreover, both UD and, to a degree, BSO, but not CB, produced binary scatterplots of cell subpopulations, with two different phagocytic/autophagic cell phenotypes. An increased number of activated cells with very high autophagy and phagocytosis was observed after UD, CB, or BSO. An interesting observation was that for quite a few, but not all cells with lower activation profile, autophagy predominated over phagocytosis. Definitely not all THP-1 cells responded to UD or BSO in a similar way, and there are several possible explanations. Firstly, it seems that autophagy might be stimulated by the accumulation of damaged proteins, heat shock proteins, or by apoptosis due to significant UD-mediated oxidative stress (Lai et al. 2016). There is experimental evidence on the crosstalk between apoptosis and autophagy (Booth et al. 2014). Secondly, monocytes may respond to the UD stimulus by differentiating into macrophages that are of two phenotypes: M1 and M2. Activated M1 macrophages produce oxygen and nitrogen radicals, secrete TNF-α and IL-1, and facilitate complement-mediated phagocytosis (Hao et al. 2012). Activated M2 macrophages express high levels of proteins that are relevant in the immune suppression and tumor progression (Mantovani et al. 2003). Thirdly, UD-induced alterations may also represent canonical and noncanonical autophagic processes, such as LC3-associated phagocytosis, but more detailed mechanistic study is required to prove the intermediacy of noncanonical autophagy in the biological effect of UD. We submit that the most suitable scenario for diversified cell activation is the host cellular protein damage, followed by increased ubiquitination, stimulation of autophagy, and phagocytosis since latex beads covered with FITC-labeled antibodies are probably unable to induce autophagy by themselves.

The present findings indicate that BSO pretreatment may induce GSH depletion and redox imbalance in THP-1 cells (data not shown) and similar results have been obtained in THP-1 cells by others (Grodzki et al. 2013). When GSH is depleted in THP-1 cells, autophagy increases, as evidenced by increased expression of LC3 protein and autophagic vesicle numbers (Mancilla et al. 2015). It has been shown that THP-1 cells are activated by PM. Once activated they release cytokines, TNF-α, superoxides, and nitric oxides (Park et al. 2016), which may be relevant to PM-induced effects, especially considering a strong prooxidative potential of UD (Manzano-Leon et al. 2013; Donaldson et al. 2001).

In the present experimental model, CB was able to normalize the BSO-induced increase of "low-positive" (autophagy/phagocytosis) cell numbers. Coarse CB should not be definitely considered as a toxic agent, but the published data show that a small-size nanomolecular carbon can trigger toxicity (Stoeger et al. 2006). However, as a nonreactive substance, CB can affect rather cell physiology than biochemistry. THP-1 cells are able to uptake CB molecules (Sahu et al. 2014), but in our experiments there was no visible competition of CB with latex beads in the phagocytosis assay. Considering CB and oxidative stress, it has been shown that fine CB is without a significant effect on redox imbalance, which demonstrates the aptitude of surface area-dependent oxidative stress (Zhao and Riediker 2014; Donaldson et al. 2003).

The present study shows a relatively small increase in the number of cells stained with LC3-associated phagocytosis after CB exposure, but the process was stimulated. Moreover, CB was able to reverse the increasing effect of BSO pretreatment on "low-positive" phagocytosis/autophagy cell. It is possible that our CB samples contained some small-size carbon since carbon nanoparticles can alter the phagocytic capacity of monocytes (Sahu et al. 2014) or can induce ubiquitin-dependent autophagy (Liu et al. 2017).

CB nanoparticles are also reported to cause cytotoxic injury, increase the level of proinflammatory chemokines, and inhibit cell growth (Yamawaki and Iwai 2006). Clinical and animal studies have also confirmed the role of CB nanoparticles in aggravating pulmonary disorders such as asthma, lung cancer, pulmonary fibrosis, and systemic cardiovascular disorders (Donaldson et al. 2005). The present findings show that the effects of CB are less affected by changes in intracellular GSH levels than those of UD and that CB molecules are significantly less disposed to induce prooxidative changes. On the other hand, there is an asymmetric synergy between BSO and UD in increasing the number of highly stimulated (autophagy and phagocytosis) cell, but this type of interaction is not visible in a fraction of cells with a lower stimulation level and with relatively higher autophagy. Since autophagy is considered an important transition from apoptosis to differentiation of monocyte (Zhang et al. 2012), the hypothesis on distinct cellular activation pathways may be functionally and diagnostically important, but it should be validated using cell sorting and by additional functional, biochemical, and cytochemical assays. It seems, however, that the ability to discriminate between these related, yet probably distinct, pathways of cell activation may be useful in the future approaches to PM-induced pneumotoxicity, inflammation, and tumorigenesis.

Conflicts of Interest The authors had no conflicts of interest to declare in relation to this article.

Ethical Approval This article does not contain any studies with human participants or animals performed by any of the authors.

References

Alexis NE, Lay JC, Zeman K, Bennett WE, Peden DB, Soukup JM, Devlin RB, Becker S (2006) Biological material on inhaled coarse fraction particulate matter activates airway phagocytes in vivo in healthy volunteers. J Allergy Clin Immunol 117:1396–1403

Anderson JO, Thundiyil JG, Stolbach A (2012) Clearing the air: a review of the effects of particulate matter air pollution on human health. J Med Toxicol 8:166–175

Baharom F, Rankin G, Blomberg A, Smed-Sörensen A (2017) Human lung mononuclear phagocytes in health and disease. Front Immunol 8:499

Bai R, Guan L, Zhang W, Xu J, Rui W, Zhang F, Ding W (2016) Comparative study of the effects of PM1-induced oxidative stress on autophagy and surfactant protein B and C expressions in lung alveolar type II epithelial MLE-12 cells. Biochim Biophys Acta 1860:2782–2792

Booth LA, Tavallai S, Hamed HA, Cruickshanks N, Denta P (2014) The role of cell signalling in the crosstalk between autophagy and apoptosis. Cell Signal 26:549–555

Chen Y, Klionsky DJ (2011) The regulation of autophagy – unanswered questions. J Cell Sci 124:161–170

Dale DC, Boxer L, Liles WC (2008) The phagocytes: neutrophils and monocytes. Blood 112:935–945

de Oliveira Alves N, Matos Loureiro AL, Dos Santos FC, Nascimento KH, Dallacort R, de Castro Vasconcellos P, de Souza Hacon S, Artaxo P, de Medeiros SR (2011) Genotoxicity and composition of particulate matter from biomass burning in the eastern Brazilian Amazon region. Ecotoxicol Environ Saf 74:1427–1433

de Oliveira Alves N, de Souza Hacon S, de Oliveira Galvao MF, Simoes Peixotoc M, Artaxo P, de Castro Vasconcellos P, de Medeiros SR (2014) Genetic damage of organic matter in the Brazilian Amazon: a comparative study between intense and moderate biomass burning. Environ Res 130:51–58

Dobrovolskaia MA, Aggarwal P, Hall JB, McNeil SE (2008) Preclinical studies to understand nanoparticle interaction with the immune system and its potential effects on nanoparticle biodistribution. Mol Pharm 5:487–495

Donaldson K, Stone V, Clouter A, Renwick L, MacNee W (2001) Ultrafine particles. Occup Environ Med 58:211–206

Donaldson K, Stone V, Borm PJ, Jimenez LA, Gilmour PS, Schins RP, Knaapen AM, Rahman I, Faux SP, Brown DM, MacNee W (2003) Oxidative stress and calcium signaling in the adverse effects of environmental particles (PM10). Free Radic Biol Med 34:1369–1382

Donaldson K, Tran L, Jimenez LA, Duffin R, Newby DE, Mills N, MacNee W, Stone V (2005) Combustion-derived nanoparticles: a review of their toxicology following inhalation exposure. Part Fibre Toxicol 2:10

Grodzki AC, Giulivi C, Lein PJ (2013) Oxygen tension modulates differentiation and primary macrophage functions in the human monocytic THP-1 cell line. PLoS One 8:e54926

Hao NB, Lu MH, Fan YH, Cao YL, Zhang ZR, Yang SM (2012) Macrophages in tumor microenvironments and the progression of tumors. Clin Dev Immunol 2012:948098

Heckmann BL, Boada-Romero E, Cunha LD, Magne J, Green DR (2017) LC3-associated phagocytosis and inflammation. J Mol Biol 429:3561–3576

Kalai Selvi S, Vinoth A, Varadharajan T, Weng CF, Vijaya Padma V (2017) Neferine augments therapeutic efficacy of cisplatin through ROS-mediated non-canonical autophagy in human lung adenocarcinoma (A549 cells). Food Chem Toxicol 103:28–40

Lai CH, Lee CN, Bai KJ, Yang YL, Chuang KJ, Wu SM, Chuanga HC (2016) Protein oxidation and degradation caused by particulate matter. Sci Rep 6:33727

Lawal AO (2017) Air particulate matter induced oxidative stress and inflammation in cardiovascular disease and atherosclerosis: the role of Nrf2 and AhR-mediated pathways. Toxicol Lett 270:88–95

Li Z, Wu Y, Chen HP, Zhu C, Dong L, Wang Y, Liu H, Xu X, Zhou J, Wu Y, Li W, Ying S, Shen H, Chen ZH (2018) MTOR suppresses environmental particle-induced inflammatory response in macrophages. J Immunol 200:2826–2834

Liu KK, Qiu WR, Naveen Raj E, Liu HF, Huang HS, Lin YW, Chang CJ, Chen TH, Chen C, Chang HC, Hwang JK, Chao JI (2017) Ubiquitin-coated nanodiamonds bind to autophagy receptors for entry into the selective autophagy pathway. Autophagy 13:187–200

Mancilla H, Maldonado R, Cereceda K, Villarroel-Espíndola F, Montes de Oca M, Angulo C, Castro MA, Slebe JC, Vera JC, Lavandero S, Concha II (2015) Glutathione depletion induces spermatogonial cell autophagy. J Cell Biochem 116:2283–2292

Mantovani A, Schioppa T, Biswas SK, Marchesi F, Allavena P, Sica A (2003) Tumor-associated macrophages and dendritic cells as prototypic type II polarized myeloid populations. Tumori 89:459–468

Manzano-Leon N, Quintana R, Sanchez B, Serrano J, Vega E, Vazquez-Lopez I, Rojas-Bracho L, Lopez-Villegas T, O'Neill MS, Vadillo-Ortega F, De Vizcaya-Ruiz A, Rosas I, Osornio-Vargas AR (2013) Variation in the composition and *in vitro* proinflammatory effect of urban particulate matter from different sites. J Biochem Mol Toxicol 27:87–97

Matsuzawa-Ishimoto Y, Hwang S, Cadwell K (2018) Autophagy and inflammation. Annu Rev Immunol 36:73–101

Nicholas SA, Sumbayev VV (2009) The involvement of hypoxia-inducible factor-1 alpha in Toll-like receptor 7/8-mediated inflammatory response. Cell Res 19:973–983

Park S, Seok JK, Kwak JY, Suh HJ, Kim YM, Boo YC (2016) Anti-inflammatory effects of pomegranate peel extract in THP-1 cells exposed to particulate matter PM10. Evid Based Complement Alternat Med 2016:6836080

Peixoto MS, de Oliveira Galvao MF, Batistuzzo de Medeiros SR (2017) Cell death pathways of particulate matter toxicity. Chemosphere 188:32–48

Sahu D, Kannan GM, Vijayaraghavan R (2014) Carbon black particle exhibits size dependent toxicity in human monocytes. Int J Inflamm 2014:827019

Singh A, Kendall SL, Campanella M (2018) Common traits spark the mitophagy/xenophagy interplay. Front Physiol 9:1172

Stoeger T, Reinhard C, Takenaka S, Schroeppel A, Karg E, Ritter B, Heyder J, Schulz H (2006) Instillation of six different ultrafine carbon particles indicates a surface area threshold dose for acute lung inflammation in mice. Environ Health Perspect 114:328–333

Szabó-Taylor KÉ, Tóth EÁ, Balogh AM, Sódar BW, Kádár L, Pálóczi K, Fekete N, Németh A,

Osteikoetxea X, Vukman KV, Holub M, Pállinger É, Nagy G, Winyard PG, Buzás EI (2017) Monocyte activation drives preservation of membrane thiols by promoting release of oxidised membrane moieties via extracellular vesicles. Free Radic Biol Med 108:56–65

Yamawaki H, Iwai N (2006) Mechanisms underlying nano-sized air-pollution-mediated progression of atherosclerosis: carbon black causes cytotoxic injury/ inflammation and inhibits cell growth in vascular endothelial cells. Circ J 70:129–140

Zhang Y, Morgan MJ, Chen K, Choksi S, Liu ZG (2012) Induction of autophagy is essential for monocyte-macrophage differentiation. Blood 119:2895–2905

Zhao J, Riediker M (2014) Detecting the oxidative reactivity of nanoparticles: a new protocol for reducing artifacts. J Nanopart Res 16:2493

Adv Exp Med Biol - Clinical and Experimental Biomedicine (2019) 4: 65–73
https://doi.org/10.1007/5584_2018_284
© Springer Nature Switzerland AG 2018
Published online: 19 October 2018

Gender-Dependent Growth and Insulin-Like Growth Factor-1 Responses to Growth Hormone Therapy in Prepubertal Growth Hormone-Deficient Children

Ewelina Witkowska–Sędek, Małgorzata Rumińska, Anna Majcher, and Beata Pyrżak

Abstract

Gender seems to be an important factor influencing the response to recombinant human growth hormone (rhGH) therapy in GH-deficient adolescents and adults. The results of studies evaluating gender-specific response to rhGH therapy in prepubertal GH-deficient children are divergent. The aim of this study was to determine the effect of gender on the growth and insulin-like growth factor-1 (IGF-1) responses in 75 prepubertal GH-deficient children during the first 2 years of rhGH therapy. There were no baseline gender differences in age, bone age, anthropometrical parameters, and IGF-1 SDS for bone age. After the initiation of rhGH therapy, there were no gender-specific differences concerning the reduction of height deficit. Serum IGF-1 levels were higher in the prepubertal GH-deficient girls than in the age-matched boys, but the difference was not significant when expressed as IGF-1 SDS for bone age. The increase in IGF-1 SDS for bone age was significantly greater in girls versus boys after the first 6 months of therapy, comparable between girls and boys after the first year of therapy, and tended to be higher in boys after the second year of therapy. In conclusion, prepubertal GH-deficient girls and boys do not differ significantly in growth response in the first 2 years of rhGH therapy.

Keywords

Body height · Gender · Growth · Growth hormone deficiency · Growth hormone therapy · Insulin-like growth factor · Prepubertal children

1 Introduction

Childhood growth hormone deficiency (GHD) was the first formally approved indication for administering growth hormone (GH), which was at first pituitary-derived and then produced with recombinant DNA technology. The discovery of recombinant human GH (rhGH) made it possible to extend the indications for this therapy. Nowadays, apart from childhood GHD, therapeutic use of rhGH includes GHD in transition from childhood to adulthood and also adult GHD (Ranke

E. Witkowska–Sędek (✉), M. Rumińska, A. Majcher, and B. Pyrżak
Department of Paediatrics and Endocrinology, Medical University of Warsaw, Warsaw, Poland
e-mail: ewelina.witkowska-sedek@wum.edu.pl

and Wit 2018). Treatment with rhGH in pediatric population seems very effective, especially in GH-deficient children whose therapy starts before puberty, but the long-term outcomes are not always satisfactory (Witkowska–Sędek et al. 2018; Witkowska–Sędek et al. 2016; Reiter et al. 2006; Lanes 2004; Wit 2002). In children with GHD, effectiveness of long-term rhGH replacement therapy depends on both pre-treatment and treatment-related factors, such as birth weight, baseline height deficit, age at the initiation of treatment, height at the start of puberty, duration of therapy, target height, and the rhGH dose and frequency of its injection (Tomaszewski et al. 2018; Polak et al. 2017; Stagi et al. 2017; Murray et al. 2016; Ross et al. 2010, 2015; Oberbauer 2014; Darendeliler et al. 2011; Cohen et al. 2002; Wasniewska et al. 2000). Gender may also be an important factor influencing the response to rhGH therapy in GH-deficient patients, but the results of available studies indicate that such associations are mainly found in GH-deficient adolescents during puberty and in GH-deficient adults, which confirms the role of sex steroids (Columb et al. 2009; Svensson et al. 2003; Drake et al. 2001; Span et al. 2000; Johansson 1999). Several studies have revealed that GH-deficient women require higher rhGH doses to achieve the same biological effects as that in men (Ezzat et al. 2002; Nilsson 2000; Bengtsson et al. 1999; Hayes et al. 1999; Burman et al. 1997). The authors evaluating gender-specific response to rhGH therapy in prepubertal GH-deficient children report divergent results, but most indicate an important role of GH doses administered (Sävendahl et al. 2012; Darendeliler et al. 2011; Ross et al. 2010; Rose et al. 2005; Ranke et al. 1999).

Gender-specific differences in the regulation of neuroendocrine control of the hypothalamic–pituitary–insulin-like growth factor-1 (IGF-1) axis and sensitivity to GH and IGF-1 are also postulated (Cohen et al. 2002; Veldhuis et al. 2000; Giustina and Veldhuis 1998). Taking into account the fact that prepubertal girls and boys differ in serum estradiol and testosterone concentrations, the content of sex steroids could lead to gender dimorphism in IGF-1 levels and

response to rhGH therapy, even before the onset of puberty (Courant et al. 2010; Demerath et al. 1999; Klein et al. 1994). The latest observations suggest that not only growth response but also metabolic outcomes of rhGH replacement therapy seem to be gender-specific in GH-deficient prepubertal children (Ciresi et al. 2018). Therefore, the aim of this study was to determine the effect of gender on the growth response and changes in IGF-1 levels in prepubertal GH-deficient children during the first 2 years of rhGH replacement therapy.

2   Methods

2.1   Patients

In this retrospective study we analyzed data of medical files of 75 prepubertal children (28 girls and 47 boys) with isolated idiopathic GHD treated with rhGH from 2011 to 2018. We examined the effect of gender on the growth response and IGF-1 levels during the first year of rhGH replacement therapy in all 75 children and during the first 2 years of therapy in 62 of them (24 girls and 38 boys). All patients remained prepubertal until the end of the study time. GHD was diagnosed based on the peak GH release below 10 ng/ml in a nocturnal test of spontaneous GH secretion and in two pharmacological tests with different stimuli (clonidine, insulin, arginine, or glucagon). All of the children fulfilled the following criteria of the Polish program for rhGH replacement therapy in short children with GHD: height below the third percentile for age and gender according to Polish growth charts, height velocity (HV) below -1 SD of the mean for age- and sex-matched Polish population, and a delay in bone age. Magnetic resonance imaging of the hypothalamic–pituitary region was conducted in all the patients to exclude organic lesions. Recombinant human GH was given subcutaneously once daily at bedtime. The mean rhGH doses (mg/kg/week) administered to girls and boys in the first 6 months and in the first and second year of rhGH replacement therapy are reported in Table 1. The following baseline and

Table 1 Comparison between growth hormone (GH)-deficient prepubertal girls and boys at baseline and during the first 2 years of recombinant human GH (rhGH) replacement therapy

Parameter	Girls	Boys	P-value
Baseline			
Number of patients	28	47	
Age (years)	7.8 ± 2.4	7.4 ± 2.1	ns
Bone age (years)	5.9 ± 2.6	5.0 ± 2.0	ns
Bone age delay (years)	1.9 ± 1.0	2.4 ± 1.1	ns
Peak GH (ng/ml)	7.7 ± 1.4	7.1 ± 2.2	ns
Height SDS	-2.58 ± 0.49	-2.71 ± 0.61	ns
Weight SDS for height–age	-0.25 ± 0.76	-0.37 ± 0.50	ns
BMI SDS for height–age	-0.40 ± 1.04	-0.47 ± 0.71	ns
HV (cm/year)	5.2 ± 1.2	4.9 ± 1.1	ns
IGF-1 (ng/ml)	108.0 (71.3–153.0)	67.7 (42.1–96.9)	< 0.01
IGF-1 SDS	0.04 (– 0.40–0.66)	– 0.25 (– 0.95–0.43)	ns
After 6 months' therapy			
Number of patients	28	47	
Height SDS	$-2.24 \pm 0.55^{***}$	$-2.28 \pm 0.56^{***}$	ns
Weight SDS for height–age	-0.35 ± 0.70	-0.36 ± 0.42	ns
BMI SDS for height–age	-0.46 ± 0.95	-0.49 ± 0.57	ns
HV (cm/6 months)	4.5 ± 0.8	5.0 ± 0.9	< 0.05
IGF-1 (ng/ml)	$249.0 (153.0–301.0)^{**}$	$142.0 (109.0–202.0)^{**}$	< 0.001
IGF-1 SDS	$2.68 (1.03–5.41)^{**}$	$1.47 (0.45–2.72)^{**}$	< 0.01
Delta IGF-1 SDS$_{\text{six months–baseline}}$	3.74 (1.60–5.58)	1.88 (0.90–2.58)	< 0.01
GH dose in the first 6 months (mg/kg/week)	0.179 ± 0.010	0.178 ± 0.010	ns
After 1 year's therapy			
Number of patients	28	47	
Age (years)	8.8 ± 2.4	8.4 ± 2.1	ns
Bone age (years)	$7.5 \pm 2.8^{***}$	$6.3 \pm 2.3^{***}$	< 0.05
Bone age delay (years)	$1.3 \pm 1.0^{***}$	2.4 ± 2.1	< 0.01
Height SDS	$-1.98 \pm 0.56^{***}$	$-2.01 \pm 0.57^{***}$	ns
Weight SDS for height–age	-0.33 ± 0.65	-0.31 ± 0.41	ns
BMI SDS for height–age	-0.42 ± 0.89	-0.35 ± 0.58	ns
HV (cm/year)	$8.6 \pm 1.2^{***}$	$9.0 \pm 1.4^{***}$	ns
IGF-1 (ng/ml)	$313.5 (140.0–377.0)^{**}$	$173.5 (131.0–221.0)^{**}$	< 0.01
IGF-1 SDS	$1.60 (0.66–4.27)^{**}$	$1.54 (0.89–2.56)^{**}$	ns
Delta IGF-1 SDS$_{\text{first year–baseline}}$	2.09 (0.85–4.08)	1.90 (0.70–3.44)	ns
GH dose in the first year (mg/kg/week)	0.179 ± 0.010	0.178 ± 0.010	ns
After 2 years' therapy			
Number of patients	24	38	
Age (years)	10.0 ± 2.3	9.5 ± 2.1	ns
Bone age (years)	$9.0 \pm 2.7^{***}$	$7.9 \pm 2.2^{***}$	ns
Bone age delay (years)	$1.0 \pm 0.9^{***}$	$1.6 \pm 1.1^{*}$	< 0.05
Height SDS	$-1.74 \pm 0.60^{***}$	$-1.66 \pm 0.6^{***}$	ns
Weight SDS for height–age	-0.30 ± 0.48	-0.27 ± 0.41	ns
BMI SDS for height–age	$-0.33 \pm 0.62^{*}$	-0.30 ± 0.55	ns
HV (cm/year)	$7.4 \pm 1.1^{***}$	$7.1 \pm 0.9^{***}$	ns
IGF-1 (ng/ml)	$316.5 (227.0–431.5)^{**}$	$217.0 (171.0–327.0)^{**}$	< 0.05
IGF-1 SDS	$2.14 (0.78–2.83)^{**}$	$2.22 (1.60–4.37)^{**}$	ns
Delta IGF-1 SDS$_{\text{second year–baseline}}$	1.88 (0.77–3.39)	2.77 (1.69–4.37)	ns
GH dose in the second year (mg/kg/week)	0.181 ± 0.01	0.177 ± 0.02	ns

Data are presented as means $\pm$ SD or median with interquartile ranges. *Peak GH* maximum growth hormone release in diagnostic tests, *SDS* standard deviation score, *BMI* body mass index, *HV* height velocity, *IGF-1* insulin-like growth factor-1, *delta IGF-1 SDS* change in IGF-1 SDS after the initiation of growth hormone replacement therapy, *p < 0.05 vs. baseline; **p < 0.01 vs. baseline; ***p < 0.001 vs. baseline

treatment parameters were evaluated: peak GH levels, height, weight, HV, serum IGF-1 concentrations, and bone age. Body mass index (BMI) was calculated as weight in kilograms divided by height in square meters. Height measurements were expressed as standard deviation scores (SDS) for chronological age, while weight measurements and BMI were expressed as SDS for height–age. Baseline HV was calculated using data from 6 to 12 months before the onset of rhGH therapy. Bone age was evaluated at baseline and after the first and the second year of rhGH therapy (Greulich and Pyle 1959).

2.2 Biochemical Analysis

Serum GH content (in pre-treatment diagnostic tests) and IGF-1 content (at baseline, after 6 months, and after the first and second year of rhGH replacement therapy) were measured by an immunoassay using Immulite 2000 Xpi Analyzer (Siemens; Erlangen, Germany). The IGF-1 content was converted to IGF-1 SDS for gender and bone age based on the normative data provided by the manufacturer (Siemens Healthcare Diagnostics Inc; Deerfield, IL).

2.3 Statistical Analysis

Data were reported as means $\pm$SD or medians and interquartile ranges. Data distribution was checked with the Shapiro–Wilk normality test. Comparisons between girls and boys were conducted using a t-test for normally distributed and the Mann–Whitney U test for non-normally distributed data. Comparisons between baseline and treatment values of the same parameter were conducted using repeated measures of ANOVA with Bonferroni post hoc test for normally distributed and the Friedman test with post hoc comparisons for non-normally distributed data. A P value <0.05 was considered significant. Data were analyzed using a Statistica v13.1 software package (StatSoft, Tulsa, OK).

3 Results

The comparison between GH-deficient prepubertal girls and boys at baseline and during the first 2 years of rhGH replacement therapy is presented in Table 1. At baseline there were no significant differences in the chronological age, bone age, peak GH response in diagnostic tests, height SDS, weight SDS for height–age, BMI SDS for height–age, or HV between girls and boys. Although there were no significant gender differences in baseline bone age delay, the tendency for greater bone age delay was observed in boys (p = 0.065). Baseline serum IGF-1 levels were significantly higher in girls than those in boys (p < 0.01), but after transforming the values to IGF-1 SDS for bone age, the difference was insignificant.

HV increased significantly after the first year of rhGH replacement therapy (p < 0.001 vs. baseline) and remained higher than baseline in the second year of rhGH therapy (p < 0.001 vs. baseline), regardless of gender. Height SDS increased significantly as early as at the first 6 months of therapy (p < 0.001 vs. baseline for both girls and boys). Weight SDS for height–age did not change significantly during rhGH therapy in either gender. BMI SDS for height–age increased significantly (p < 0.05) in girls after the second year of therapy compared to the baseline values and tended to increase in the same period in boys, without reaching statistical significance. Bone age advanced significantly after the onset of rhGH therapy in both girls and boys (p < 0.001 vs. baseline). The serum IGF-1 content (p < 0.01 for both genders) and IGF-1 SDS for bone age (p < 0.01 for both genders) increased significantly as early as at the first 6 months of therapy and remained higher than baseline throughout the 2-year treatment period. All the significant changes in anthropometric and biochemical parameters during rhGH therapy compared to the baseline values are reported in Table 1.

There were no significant gender differences in height SDS, weight SDS for height–age, and BMI SDS for height–age during rhGH therapy. Height velocity in the first 6 months of therapy was significantly higher in boys than in girls (p < 0.05), but when calculated after the first and the second year of therapy, it failed to differ significantly between genders. Bone age was significantly higher in girls (p < 0.05) than that in boys after the first year and tended to be higher in girls after the second year of therapy as well (p = 0.071). The serum IGF-1 content remained significantly higher in girls than that in boys after the first 6 months (p < 0.001) and after the first (p < 0.01) and second (p < 0.05) year of therapy, but IGF-1 SDS for bone age was significantly higher in girls than that in boys only after the first 6 months of therapy (p < 0.01).

There were no significant gender-specific differences in the reduction in height deficit (delta height SDS) after the first 6 months, the first year, and the second year of rhGH therapy (data not shown). A significant reduction in bone age delay compared to the baseline values was observed after the initiation of therapy in both genders, but in girls it occurred as early as after the first year of therapy (p < 0.001), and in boys later, after the second year of therapy (p < 0.05). Despite that, bone age delay remained significantly higher in boys than that in girls after both the first (p < 0.01) and the second (p < 0.05) years of rhGH therapy.

The increase in IGF-1 SDS for bone age after the initiation of rhGH therapy, in comparison to the baseline values (delta IGF-1 SDS), was significantly higher in girls than that in boys after the first 6 months (IGF-1 $SDS_{six\ months-baseline}$) of therapy (p < 0.01), did not differ significantly after the first year (IGF-1 $SDS_{first\ year-baseline}$), and tended to be higher in boys after the second year (IGF-1 $SDS_{second\ year-baseline}$) of therapy (p = 0.069). The comparison of changes in IGF-1 SDS during rhGH replacement therapy between girls and boys is presented in Fig. 1.

4   Discussion

In this study we failed to find any significant differences in height deficit, weight SDS for height–age, and BMI SDS for height–age between age-matched prepubertal girls and boys

with isolated idiopathic GHD either at baseline or during the first 2 years of rhGH therapy. Baseline height velocity, bone age, and peak GH levels in diagnostic tests did not differ significantly between genders either. Although height velocity in the first 6 months of therapy was higher in boys than in girls, height deficit reduction during that period was not different between genders. There were no gender differences in height velocity calculated after the first year of rhGH therapy or in height velocity in the second year of therapy. Our findings were similar to those reported earlier by other researchers. Rose et al. (2005), who analyzed data of prepubertal children with isolated idiopathic GHD from the Pfizer Kabi International Growth Study (KIGS) database, did not find any statistically significant gender differences in height gain and height velocity after the second and third year of rhGH replacement therapy. In contrast to our observations, those authors have noticed that baseline height deficit is greater in girls than in boys, and the difference remains significant after the initiation of rhGH therapy. The authors postulate that these baseline differences in height deficit could reflect greater societal tolerance for short stature in girls than in boys. They conclude that their results support the hypothesis that gender differences in response to rhGH replacement therapy observed in GH-deficient patients, mainly adolescents and adults, result from the effects of sex steroids on tissues and binding proteins rather than from chromosomal differences, and that is why the differences are not apparent before puberty. The limitations of their study are that IGF-1 content was not analyzed at baseline and after the initiation of rhGH therapy, and that a narrow range of rhGH dose was used (0.25–0.35 mg/kg/week). A later study by Ross et al. (2010) has revealed that there were no gender-specific differences during the 2-year-long rhGH therapy not only in GH-deficient children but also in other groups of children treated with rhGH, such as children with idiopathic short stature, multiple pituitary hormone deficiency, or Noonan syndrome. Similar results of short- and long-term rhGH therapy have been reported in an earlier study of Haffner et al. (1998) in a group of prepubertal children with

Fig. 1 Comparison of changes (delta (Δ) difference) in insulin-like growth factor-1 standard deviation score (IGF-1 SDS) between girls and boys after 6 months, the first year, and the second year of recombinant human growth hormone (rhGH) replacement therapy. The difference between boys and girls was significant after 6 months of therapy (p < 0.05), insignificant after 1 year of therapy, and tended to be of borderline significance after 2 years of therapy (p = 0.069)

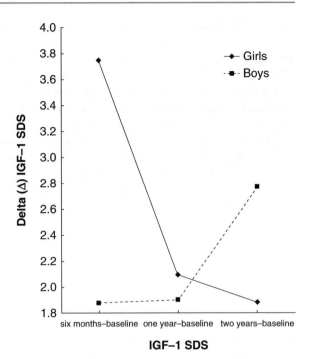

chronic renal failure. In contrast to the abovementioned studies, Sävendahl et al. (2012) have reported a significantly higher reduction in height deficit in prepubertal GH-deficient and small for gestational age boys versus girls during the first 2 years of rhGH therapy, with no gender differences in changes of IGF-1 SDS.

In the present study, we found that the serum baseline IGF-1 levels were significantly higher in girls than in boys; the gender difference was sustained after 6 months and the first and second year of rhGH replacement therapy. This difference, however, was unaccompanied by other gender differences in chronological age, height deficit, or peak GH level in diagnostic tests. Given that our study group consisted of prepubertal children of the mean age not exceeding 8 years, the results are in opposition to the fact that in healthy prepubertal children the IGF-1 levels do not differ between genders and depend only on age (Bereket et al. 2006; Löfqvist et al. 2001). Significant gender-specific differences in the serum IGF-1 level are observed after the onset of puberty and reach the highest values earlier in girls than in boys (14 vs.

15 years) (Bereket et al. 2006; Löfqvist et al. 2001; Juul et al. 1994). On the other hand, Yüksel et al. (2011), who analyzed data of a large cohort of healthy children younger than 6 years of age, confirmed that serum IGF-1 levels slowly increase with age but simultaneously found that IGF-1 levels differed between genders, and were lower in girls than in boys only at the age of 6 months. In the present study, dissimilarities in serum IGF-1 concentrations were observed at baseline and during rhGH therapy, but differences in IGF-1 SDS for bone age were found only after the first 6 months of rhGH therapy, which could reflect the gender differences in bone age. The delay in bone age tended to be higher in boys than in girls at baseline, and then it became significantly higher in boys versus girls after the first and second year of rhGH therapy. The increase in IGF-1 SDS during rhGH therapy was significantly higher in girls versus boys during the first 6 months of therapy. The increase then did not differ significantly when calculated after the first year of therapy and tended to be higher in boys than in girls within the first 2 years of therapy. Although the higher increase in IGF-1 SDS

within the first 6 months of therapy was found in girls, height velocity in that period was greater in boys. Those findings seem to support the hypothesis that sensitivity to IGF-1 and responsiveness to rhGH are gender-specific, even in prepubertal children (Cohen et al. 2002), and that prepubertal GH-deficient girls require higher IGF-1 levels to achieve similar growth response to that of GH-deficient prepubertal boys. Previous studies, which indicate gender dimorphism concerning the content of sex steroids, not only after the onset of puberty but also in prepubertal children (Courant et al. 2010; Klein et al. 1994), could partially explain gender-specific differences in IGF-1 sensitivity and response to rhGH therapy. Estradiol levels in prepubertal girls are much higher than in prepubertal boys (Courant et al. 2010; Klein et al. 1994), and conversely, serum testosterone is higher in prepubertal boys than in prepubertal girls (Demerath et al. 1999). Differences in body composition, reported in prepubertal girls and boys, could also contribute to gender-specific IGF-1 and GH responsiveness (Abdenur et al. 1994).

The important role of GH dose administered to prepubertal GH-deficient children during therapy is worthy of note. Cohen et al. (2002) have shown that the effects of rhGH therapy on growth and serum IGF-1 levels are gender-specific, only in patients who receive sufficiently high rhGH doses. Those authors randomized prepubertal GH-deficient children to receive low (0.025 mg/kg per day), medium (0.05 mg/kg per day), and high (0.10 mg/kg per day) rhGH dose and found that the increases in height SDS and serum IGF-1 and IGF-binding protein-3 (IGFBP-3) were dose- and gender-specific. Boys had a linear GH dose response, whereas girls had an apparent plateau of both growth and IGF-1 SDS responses at medium rhGH dose. The authors suggest that the efficacy and safety of GH therapy can be optimized by modulating the rhGH dose in a gender-specific manner, based on the growth response and serum IGF-1 and IGFBP-3 levels. All of the patients of the present study received an rhGH dose, on average, about 0.18 mg/kg/week, which

approximately corresponds to the lower limit of the recommended range. We did notice, even with this small dose, a plateau in the IGF-1 SDS increase of the first 6 months of therapy, after the further first and second year of therapy.

Ciresi et al. (2018), who have evaluated the effect of gender on several clinical and glucolipid metabolism parameters in a cohort of prepubertal GH-deficient children during the first 2 years of rhGH replacement therapy, report a lack of gender-specific differences in height, IGF-1 values, fasting glucose and insulin levels, glycated hemoglobin (HbA1c), insulin resistance index (HOMA–IR), and the lipid profile. On the other hand, those authors have found that after the second year of therapy, girls show a significantly higher BMI and lower insulin sensitivity index (ISI) and oral disposition index (DIo) than boys. Taking into account that IGF-1 is involved in the glucose metabolism alterations during rhGH therapy (Witkowska–Sędek et al. 2018; Rothermel and Reinehr 2016; Vijayakumar et al. 2010), higher IGF-1 content found in girls in the present study supports the hypothesis that girls are at a higher risk of glucose-related abnormalities. The relationships between gender and metabolic profile during rhGH therapy require further studies using alternative study designs.

In conclusion, prepubertal growth hormone-deficient girls and boys do not differ significantly in growth response during the first 2 years of rhGH replacement therapy. Baseline and on-treatment content of serum insulin-like growth factor-1 is higher in prepubertal growth hormone-deficient girls than that in age-matched boys, but the difference becomes insignificant when expressed as IGF-1 SDS for bone age at baseline and after the first and second year of rhGH therapy.

Conflicts of Interest The authors declare no conflict of interest in relation to this article.

Ethical Approval All procedures performed in the study were in accordance with the ethical standards of the institutional research committee and with the 1964 Helsinki declaration and its later amendments. The study was approved by the Bioethics Committee of the Medical University of Warsaw, Poland.

Informed Consent Informed written consent was obtained from the parents or legal guardians of all individual participants included in the study.

References

Abdenur JE, Solans CV, Smith MM, Carman C, Pugliese MT, Lifshitz F (1994) Body composition and spontaneous growth hormone secretion in normal short stature children. J Clin Endocrinol Metab 78(2):277–282

Bengtsson BA, Abs R, Bennmarker H, Monson JP, Feldt–Rasmussen U, Hernberg–Stahl E, Westberg B, Wilton P, Wüster C (1999) The effects of treatment and the individual responsiveness to growth hormone (GH) replacement therapy in 665 GH–deficient adults. KIMS Study Group and the KIMS International Board. J Clin Endocrinol Metab 84(11):3929–3935

Bereket A, Turan S, Omar A, Berber M, Ozen A, Akbenlioglu C, Haklar G (2006) Serum IGF–I and IGFBP–3 levels of Turkish children during childhood and adolescence: establishment of reference ranges with emphasis on puberty. Horm Res 65(2):96–105

Burman P, Johansson AG, Siegbahn A, Vessby B, Karlsson FA (1997) Growth hormone (GH)–deficient men are more responsive to GH replacement therapy than GH–deficient women. J Clin Endocrinol Metab 82(2):550–555

Ciresi A, Radellini S, Guarnotta V, Mineo MG, Giordano C (2018) The metabolic outcomes of growth hormone treatment in children are gender–specific. Endocr Connect 7(7):879–887

Cohen P, Bright GM, Rogol AD, Kappelgaard AM, Rosenfeld RG, American Norditropin Clinical Trials Group (2002) Effects of dose and gender on the growth and growth factor response to GH in GH–deficient children: implications for efficacy and safety. J Clin Endocrinol Metab 87(1):90–98

Columb B, Smethurst LE, Mukherjee A, Jostel A, Shalet SM, Murray RD (2009) GH sensitivity of GH–deficient adults is dependent on gender but not timing of onset. Clin Endocrinol 70(2):281–286

Courant F, Aksglaede L, Antignac JP, Monteau F, Sorensen K, Andersson AM, Skakkebaek NE, Juul A, Bizec BL (2010) Assessment of circulating sex steroid levels in prepubertal and pubertal boys and girls by a novel ultrasensitive gas chromatography–tandem mass spectrometry method. J Clin Endocrinol Metab 95(1):82–92

Darendeliler F, Lindberg A, Wilton P (2011) Response to growth hormone treatment in isolated growth hormone deficiency versus multiple pituitary hormone deficiency. Horm Res Paediatr 76(Suppl 1):42–46

Demerath EW, Towne B, Wisemandle W, Blangero J, Chumlea WC, Siervogel RM (1999) Serum leptin concentration, body composition, and gonadal hormones during puberty. Int J Obes Relat Metab Disord 23(7):678–685

Drake WM, Rodríguez–Arnao J, Weaver JU, James IT, Coyte D, Spector TD, Besser GM, Monson JP (2001) The influence of gender on the short and long–term effects of growth hormone replacement on bone metabolism and bone mineral density in hypopituitary adults: a 5–year study. Clin Endocrinol 54(4):525–532

Ezzat S, Fear S, Gaillard RC, Gayle C, Landy H, Marcovitz S, Mattioni T, Nussey S, Rees A, Svanberg E (2002) Gender–specific responses of lean body composition and non–gender–specific cardiac function improvement after GH replacement in GH–deficient adults. J Clin Endocrinol Metab 87(6):2725–2733

Giustina A, Veldhuis JD (1998) Pathophysiology of the neuroregulation of growth hormone secretion in experimental animals and the human. Endocr Rev 19(6):717–797

Greulich WW, Pyle SI (1959) Radiographic atlas of skeletal development of the hand and wrist, 2nd edn. Stanford University Press, Stanford, CA

Haffner D, Wühl E, Schaefer F, Nissel R, Tönshoff B, Mehls O (1998) Factors predictive of the short– and long–term efficacy of growth hormone treatment in prepubertal children with chronic renal failure. The German Study Group for Growth Hormone Treatment in Chronic Renal Failure. J Am Soc Nephrol 9(10):1899–1907

Hayes FJ, Fiad TM, McKenna TJ (1999) Gender difference in the response of growth hormone (GH)–deficient adults to GH therapy. Metabolism 48(3):308–313

Johansson AG (1999) Gender difference in growth hormone response in adults. J Endocrinol Investig 22(Suppl 5):58–60

Juul A, Bang P, Hertel NT, Main K, Dalgaard P, Jørgensen K, Müller J, Hall K, Skakkebaek NE (1994) Serum insulin–like growth factor–1 in 1030 healthy children, adolescents, and adults: relation to age, sex, stage of puberty, testicular size, and body mass index. J Clin Endocrinol Metab 78(3):744–752

Klein KO, Baron J, Colli MJ, McDonnell DP, Cutler GB Jr (1994) Estrogen levels in childhood determined by an ultrasensitive recombinant cell bioassay. J Clin Invest 94(6):2475–2480

Lanes R (2004) Long–term outcome of growth hormone therapy in children and adolescents. Treat Endocrinol 3(1):53–66

Löfqvist C, Andersson E, Gelander L, Rosberg S, Blum WF, Albertsson Wikland K (2001) Reference values for IGF–I throughout childhood and adolescence: a model that accounts simultaneously for the effect of gender, age, and puberty. J Clin Endocrinol Metab 86(12):5870–5876

Murray PG, Dattani MT, Clayton PE (2016) Controversies in the diagnosis and management of growth hormone deficiency in childhood and adolescence. Arch Dis Child 101(1):96–100

Nilsson AG (2000) Effects of growth hormone replacement therapy on bone markers and bone mineral density in growth hormone–deficient adults. Horm Res 54(Suppl 1):52–57

Oberbauer AM (2014) The influence of growth hormone on bone and adipose programming. Adv Exp Med Biol 814:169–176

Polak M, Blair J, Kotnik P, Pournara E, Pedersen BT, Rohrer TR (2017) Early growth hormone treatment start in childhood growth hormone deficiency improves near adult height: analysis from NordiNet® International Outcome Study. Eur J Endocrinol 177 (5):421–429

Ranke MB, Wit JM (2018) Growth hormone - past, present and future. Nat Rev Endocrinol 14(5):285–300

Ranke MB, Lindberg A, Chatelain P, Wilton P, Cutfield W, Albertsson–Wikland K, Price DA (1999) Derivation and validation of a mathematical model for predicting the response to exogenous recombinant human growth hormone (GH) in prepubertal children with idiopathic GH deficiency. KIGS International Board. Kabi Pharmacia International Growth Study. J Clin Endocrinol Metab 84(4):1174–1183

Reiter EO, Price DA, Wilton P, Albertsson–Wikland K, Ranke MB (2006) Effect of growth hormone (GH) treatment on the near–final height of 1258 patients with idiopathic GH deficiency: analysis of a large international database. J Clin Endocrinol Metab 91(6):2047–2054

Rose SR, Shulman DI, Larsson P, Wakley LR, Wills S, Bakker B (2005) Gender does not influence prepubertal growth velocity during standard growth hormone therapy - analysis of United States KIGS data. J Pediatr Endocrinol Metab 18(11):1045–1051

Ross J, Lee PA, Gut R, Germak J (2010) Factors influencing the one– and two–year growth response in children treated with growth hormone: analysis from an observational study. Int J Pediatr Endocrinol 2010:494656

Ross JL, Lee PA, Gut R, Germak J (2015) Increased height standard deviation scores in response to growth hormone therapy to near–adult height in older children with delayed skeletal maturation: results from the ANSWER Program. Int J Pediatr Encorinol 2015(1):1

Rothermel J, Reinehr T (2016) Metabolic alterations in paediatric GH deficiency. Best Pract Res Clin Endocrinol Metab 30(6):757–770

Sävendahl L, Blankenstein O, Oliver I, Christesen HT, Lee P, Pedersen BT, Rakov V, Ross J (2012) Gender influences short–term growth hormone treatment response in children. Horm Res Paediatr 77 (3):188–194

Span JP, Pieters GF, Sweep CG, Hermus AR, Smals AG (2000) Gender difference in insulin–like growth factor I response to growth hormone (GH) treatment in GH–deficient adults: role of sex hormone replacement. J Clin Endocrinol Metab 85(3):1121–1125

Stagi S, Scalini P, Farello G, Verrotti A (2017) Possible effects of an early diagnosis and treatment in patients with growth hormone deficiency: the state of art. Ital J Pediatr 43(1):81

Svensson J, Sunnerhagen KS, Johannsson G (2003) Five years of growth hormone replacement therapy in adults: age– and gender–related changes in isometric and isokinetic muscle strength. J Clin Endocrinol Metab 88(5):2061–2069

Tomaszewski P, Milde K, Majcher A, Pyrżak B, Tiryaki–Sonmez G, Schoenfeld BJ (2018) Body mass disorders in healthy short children and in children with growth hormone deficiency. Adv Exp Med Biol 1023:55–63

Veldhuis JD, Roemmich JN, Rogol AD (2000) Gender and sexual maturation–dependent contrasts in the neuroregulation of growth hormone secretion in prepubertal and late adolescent males and females - a general clinical research center–based study. J Clin Endocrinol Metab 85(7):2385–2394

Vijayakumar A, Novosyadlyy R, Wu YJ, Yakar S, LeRoith D (2010) Biological effects of growth hormone on carbohydrate and lipid metabolism. Growth Hormon IGF Res 20:1–7

Wasniewska M, Arrigo T, Cisternino M, De Luca F, Ghizzoni L, Maghnie M, Valenzise M (2000) Birth weight influences long–term catch–up growth and height prognosis of GH–deficient children treated before the age of 2 years. Eur J Endocrinol 142 (5):460–465

Wit JM (2002) Growth hormone therapy. Best Pract Res Clin Endocrinol Metab 16(3):483–503

Witkowska–Sędek E, Kucharska A, Rumińska M, Pyrżak B (2016) Relationship between 25(OH)D and IGF–I in children and adolescents with growth hormone deficiency. Adv Exp Med Biol 912:43–49

Witkowska–Sędek E, Labochka D, Stelmaszczyk–Emmel A, Majcher A, Kucharska A, Sobol M, Kadziela K, Pyrzak B (2018) Evaluation of glucose metabolism in children with growth hormone deficiency during long–term growth hormone treatment. J Physiol Pharmacol 69(2). https://doi.org/10.26402/jpp.2018.2.08

Yüksel B, Özbek MN, Mungan NÖ, Darendeliler F, Budan B, Bideci A, Çetinkaya E, Berberoğlu M, Evliyaoğlu O, Yeşilkaya E, Arslanoğlu İ, Darcan Ş, Bundak R, Ercan O (2011) Serum IGF–1 and IGFBP–3 levels in healthy children between 0 and 6 years of age. J Clin Res Pediatr Endocrinol 3(2):84–88

Adv Exp Med Biol - Clinical and Experimental Biomedicine (2019) 4: 75–81
https://doi.org/10.1007/5584_2018_310
© Springer Nature Switzerland AG 2019
Published online: 12 January 2019

Vitamin D and Calcium Homeostasis in Infants with Urolithiasis

Agnieszka Szmigielska, Małgorzata Pańczyk-Tomaszewska, Małgorzata Borowiec, Urszula Demkow, and Grażyna Krzemień

Abstract

The incidence of urolithiasis in infants is unknown. The aim of this study was to investigate clinical characteristics, nutrition, calcium, phosphate, 25-hydroxyvitamin D (25 (OH)D), alkaline phosphate, and parathyroid hormone in infants with urolithiasis. There were 32 infants (23 boys and 9 girls) of the mean age of 6.4 ± 3.7 months (range 2–12 months), with diagnosis of urolithiasis enrolled into the study. Boys were younger than girls (5.3 vs. 9.1 months, respectively; $p < 0.05$). The infants were receiving prophylactic vitamin D_3. Twenty-one of them were fed with milk formula, 9 were breastfed, and 2 were on a mixed diet. The major clinical symptoms consisted of irritability in 19 (59%) and urinary tract infection in 6 (19%) infants. Hypercalcemia and hyperphosphatemia were detected in the serum in 30 (94%) and 19 (60%) infants, respectively. The serum calcium level was higher in boys than girls (10.8 vs. 9.8 mg/dL, respectively; $p < 0.05$). Four (12.5%) infants had increased activity of alkaline phosphatase. The serum level of 25(OH)D was high in 3 (9%), low in 2 (6%), and normal in 27 (85%) infants. Parathyroid hormone was low in eight (25%) infants. Hypercalciuria and hyperphosphaturia were found in 11 (34%) boys and 8 (25%) girls. Family history of urolithiasis was positive in eight (25%) infants. We conclude that urolithiasis occurs in infancy more often in boys fed with milk formula and in those who received vitamin D supplementation. Hypercalcemia, hyperphosphatemia, and hypercalciuria are the most common changes present in clinical metabolic tests.

A. Szmigielska (✉), M. Pańczyk-Tomaszewska, and G. Krzemień
Department of Pediatrics and Nephrology, Warsaw Medical University, Warsaw, Poland
e-mail: aszmigielska@wum.edu.pl

M. Borowiec
Student Scientific Group at the Department of Pediatrics and Nephrology, Warsaw Medical University, Warsaw, Poland

U. Demkow
Department of Laboratory Diagnostics and Clinical Immunology of Developmental Age, Warsaw Medical University, Warsaw, Poland

Keywords

Calcium homeostasis · Hypercalcemia · Hypercalciuria · Hyperphosphaturia · Infants · Urolithiasis · Vitamin D_3

1 Introduction

The incidence of pediatric urolithiasis depends on the geographic, genetic, and socioeconomic factors and is on the rise worldwide (Weigert and Hoppe 2018; Jobs et al. 2014; López and Hoppe 2010). Approximately 10% of all cases of urinary stones are diagnosed in infants (Hesse 2005). Urolithiasis appears at any age in pediatric population. In the youngest infants, predisposing causes can be recognized in 75–80% of patients (Walther et al. 1980). The most common risk factors of urinary stone formation are metabolic changes such as hypercalciuria, hyperoxaluria, and hypocitraturia (Milliner and Murphy 1993).

25-Hydroxyvitamin D (25(OH)D) has pleiotropic effects in human body, since vitamin D receptors are expressed in the majority of cells (Unal et al. 2014). The all-presence of vitamin D_3 receptors has clinical implications in that it may have to do with a reduction of risk of cancer, autoimmune and infectious diseases, depression, diabetes, and cardiovascular incidents (Pludowski et al. 2018). Vitamin D_3 in a dose of 400 IU a day for infants, who are exclusively breastfed, is recommended by the Institute of Medicine of the US National Academy of Sciences, the American Endocrine Society, the European Society for Pediatric Gastroenterology Hepatology and Nutrition, and the American Academy of Pediatrics to prevent rickets (Holick et al. 2011; IOM 2011). The serum concentration of 25(OH)D between 20 and 50 ng/mL is established as the normal range. A routine measurement of 25(OH)D is not recommended for healthy infants, but those with chronic diseases can benefit from monitoring. In some patients, however, vitamin D_3 supplementation may have significant adverse effects such as hypercalcemia, hypercalciuria, constipation, hypertension, nephrocalcinosis, or urolithiasis. Therefore, this study seeks to define the relationship between the content of 25(OH)D and calcium and urolithiasis in infants aged under 12 months.

2   Methods

Thirty-two infants aged 2–12 months (23 boys and 9 girls), who had been referred to the Department of Pediatrics and Nephrology of Warsaw Medical University in Warsaw, Poland, due to newly diagnosed urolithiasis, were investigated. The exclusion criteria were chronic liver or kidney diseases, history of prematurity or low birth weight (< 2,500 g), and a failure to growth. The survey was conducted on nutrition, vitamin D_3 supplementation, clinical symptoms, and family history of urolithiasis. The serum levels of calcium (Ca), phosphate (P), alkaline phosphatase (ALP), parathyroid hormone (PTH), 25(OH)D, and capillary blood gas and acid-base content were investigated. In addition, urinary calcium, phosphate, magnesium, and creatinine were investigated in random urine samples. These measurements were conducted with a dry-chemistry method (VITROS 5600, Ortho Clinical Diagnostics, Raritan, NJ 08869). The PTH content was measured with an immunoenzymatic method using the Immulite 2000xPi system (Siemens Medical Solutions Diagnostics, NJ) with a reference range of 10–65 pg/mL. The serum content of vitamin D_3 was measured with the chemiluminescence method using the ARCHITECT i1000SR system (Abbott Diagnostics; Chicago, IL), with a reference range of 20–50 ng/mL. The urine calcium/creatinine (Ca/Cr), magnesium/creatinine, and phosphate/creatinine (P/Cr) ratios were calculated. The reference range for Ca/Cr was taken as less than 0.81 (Habbig et al. 2011; Hoppe and Kemper 2010). The infants were divided into the female and male groups. Metabolic indicators were compared between the two groups.

Data were presented as means ±SD and 95 confidence intervals (95%CI). Differences between the female and male groups were evaluated with a two-tailed unpaired t-test. A p-value <0.05 defined statistically significant differences. The analysis was performed using a commercial Statistica package v11.0 for Windows (StatSoft; Tulsa, OK).

3   Results

Table 1 shows the baseline characteristics of infants with urolithiasis. Urinary stones were diagnosed by means of abdominal ultrasonography. The calculi were seen in the left kidney in

Table 1 Baseline demographic and nutritional features of infants with urolithiasis

Mean age (months)	6.4 ± 3.7
Birth weight (g)	3,260–4,020
Males; n (%)	23 (72)
Females; n (%)	9 (28)
Milk formula feeding; n (%)	21 (66)
Breastfeeding; n (%)	9 (28)
Vitamin D_3 supplementation; n (%)	32 (100)
Family history of urolithiasis; n (%)	8 (25)

25 (78%) infants, in 17 (53%) in the right kidney, and in 9 (28%) bilaterally. Ultrasonography was ordered in infants due to unexplained irritability in 19 (59%), urinary tract infection in 6 (19%), or a dilation of the urinary collecting system in 7 (22%) cases. The urinary tract infection was caused by *Pseudomonas aeruginosa* in one, *Klebsiella* spp. in four, and *Enterobacter* spp. in one infant. All of the infants received oral vitamin D_3 in a total dose of 400–500 IU a day since the neonatal period. Twenty-one infants received only milk formula, supplemented with 40–56 IU of vitamin D_3 *per* 100 ml of milk, 9 were breastfed, and 2 were both breastfed and also received milk formula. A family history of urolithiasis was negative in 24 (75%) infants (Table 1).

Hypercalcemia was detected in 30 (95%) and hyperphosphatemia in 19 (60%) infants with urolithiasis. The serum PTH content was below 10 pg/mL in 8 (25%) infants, and there was no single case of increased PTH. An increased level of ALP was noticed in only 4 (12%) infants, with the remaining 28 infants having a normal ALP level. Blood gas and acid-base content was within the norm in all infants. The serum 25(OH)D content was elevated in three (9%) and was below the lower cutoff value in two (6%) infants with urolithiasis. The remaining 27 infants had the level of 25(OH)D within the normal range (Fig. 1).

All of the infants had a fresh urine sample examined under light microscopy to exclude cystinuria. Hypercalciuria was found in 11 (34.4%) patients with urolithiasis; the excretion of calcium was within the norm in 21 (65.6%) infants. Hyperphosphaturia was found in eight (25%) infants. Eighteen (56.2%) infants had the Mg/Cr ratio > 0.2 (Table 2).

The results also were stratified into the female (mean age 9.1 ± 3.3 months) and male (mean age 5.3 ± 2.4 months) groups. The only significant gender difference found was a higher calcium level in the male infants (Table 3). In addition, we found a significant inverse correlation between calcium and both 25(OH)D and ALP (Table 4).

4 Discussion

Urolithiasis affects infants of all ages. In a large study done by Zafar et al. (2018), boys are more frequently affected, during the first 2 years of life. The current study provided similar results as we noticed that there were 72% of males and 28% of females in the investigated population of infants with urolithiasis. However, symptoms of urolithiasis in infants are non-specific and can remain asymptomatic for a long time. Therefore, the exact incidence of urinary tract stones is largely unknown. In a study of Marzuillo et al. (2017), about half or more infants with urolithiasis presented abdominal or flank pain without specific urinary symptoms. In the current study, the most common symptom of urolithiasis was unexplained irritability detected in 59.4% of infants. In fact, colicky pain, hematuria, sterile pyuria, flank tenderness, and urinary retention are rather atypical symptoms for the age of few months. A routine ultrasonography in a patient with urinary tract infection is helpful in the diagnosis of the underlying urinary tract stones. In the current study, the etiology of urinary tract infections in infants with urolithiasis was atypical, consisting mostly of *Pseudomonas aeruginosa*, *Klebsiella* spp., and *Enterobacter* spp.

In the clinical setting, a reliable medical history is extremely important for the appropriate diagnosis and treatment of urolithiasis in infants. Approximately 40% of infants with urinary tract stones have a positive family history (Van't Hoff 2004). In the infants of the present, 25% of them had a family history of urolithiasis. We observed that the most important information from medical history is that on nutrition and fluid intake. Although breastfeeding in the first year of life is

Table 2 Biochemical tests in infants with urolithiasis

Biochemical tests		Reference range
Serum		
Ca; mg/dL	10.5 ± 1.1 (10.1–10.9)	8.5–10.2
P; mEq/L	4.5 ± 1.9 (3.8–5.1)	2.3–3.8
PTH; pg/mL	17.4 ± 11.6 (12.9–21.8)	10–65
25(OH)D; ng/mL	35 ± 18 (28–42)	20–50
ALP; IU/L	244 ± 70 (218–270)	91–258
Urine		
Ca/Cr; mg/mg	0.50 ± 0.27 (0.40–0.60)	<0.81
P/Cr; mg/mg	1.18 ± 0.76 (0.90–1.47)	0.30–1.20
Mg/Cr; mg/mg	0.28 ± 0.35 (0.14–0.42)	0.10

Data are means ±SD (95%CI); *Ca* calcium, *P* phosphorus, *PTH* parathyroid hormone, *25(OH)D* 25-hydroxyvitamin D, *ALP* alkaline phosphatase, *Cr* creatinine, *Mg* magnesium

Fig. 1 Serum 25(OH)D content in infants with urolithiasis

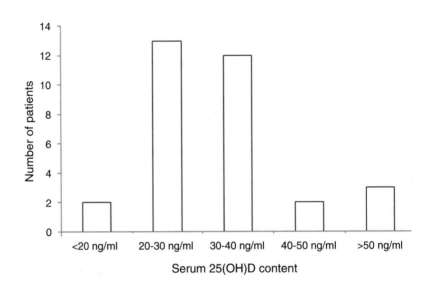

Table 3 Biochemical tests infants with urolithiasis stratified by gender

Biochemical tests	Female infants (n=9)	Male infants (n=23)	Reference range
Serum			
Ca; mg/dL	9.8 ± 1.8	10.8 ± 0.4*	8.5–10.2
P; mEq/L	3.96 ± 0.81	4.61 ± 2.11	2.25–3.78
PTH; pg/mL	21.1 ± 9.9	16.2 ± 12.1	10–65
25(OH)D; ng/mL	45 ± 30	32 ± 12	20–50
ALP; U/L	218 ± 61	256 ± 72	91–258
Urine			
Ca/Cr; mg/mg	0.42 ± 0.30	0.53 ± 0.26	<0.81
P/Cr; mg/mg	1.24 ± 0.77	1.16 ± 0.77	0.30–1.20
Mg/Cr; mg/mg	0.23 ± 0.12	0.30 ± 0.42	0.10

Data are means ±SD; *Ca* calcium, *P* phosphorus, *PTH* parathyroid hormone, *25(OH)D* 25-hydroxyvitamin D, *ALP* alkaline phosphatase, *Cr* creatinine, *Mg* magnesium; *$p < 0.05$

Table 4 Correlation matrix showing the mutual correlations among the biochemical variables investigated in infants with urolithiasis

	Ca	P	PTH	25OHD	ALP	Ca/Cr	P/Cr	Mg/Cr
Ca	1.00	−0.07	0.06	−0.70**	0.26	−0.08	0.30	0.00
P		1.00	−0.04	0.10	−0.07	0.13	−0.03	−0.02
PTH			1.00	−0.16	−0.12	−0.22	0.14	−0.11
25(OH)D				1.00	−0.45	0.08	−0.01	−0.08
ALP					1.00	0.21	0.03	0.04
Ca/Cr						1.00	0.16	0.07
P/Cr							1.00	0.04
Mg/Cr								1.00

Ca calcium, *P* phosphorus, *PTH* parathyroid hormone, *25(OH)D* 25-hydroxyvitamin D, *ALP* alkaline phosphatase, *Cr* creatinine, *Mg* magnesium
$**p < 0.001$

commonly recommended, most of the infants studied (66%) received milk formula, which is fortified with vitamin D_3. Yet aside from that, each infant was supplemented with an additional dose of 400–500 IU vitamin D daily.

It ought to be recognized that urolithiasis is only a symptom of other underlying diseases. Infants with urinary tract stones are more likely to have metabolic disorders, especially hypercalciuria and hypocitraturia. The prevalence of hypercalciuria in healthy infants has been reported at 3–7% (Skalova and Lutilek 2006). However, the incidence of hypercalciuria in patients with urolithiasis is much higher. In the current study, hypercalciuria was found in 34.4% of patients. Similar observation has been done in a study of Ergon et al. (2017) in which hypercalciuria is observed in 38% of infants with urolithiasis. Fallahzadeh et al. (2012) have evaluated 36 infants with urolithiasis and found the presence of hypercalciuria in 27.8% of them.

Hypercalciuria can be primary or secondary. Primary idiopathic hypercalciuria is the most common cause of calcium-containing stones in infants. Hypercalciuria and metabolic alkalosis also are observed in a number of genetic diseases such as Bartter's syndrome type 1, type 2, or infantile type with sensorineural deafness. Other genetic causes of hypercalciuria with urolithiasis are autosomal dominant hypocalcemic hypercalciuria (ADHH) (OMIM Entry – *601199) due to mutations in the calcium-sensing receptor (CASR) gene, familial hypomagnesemia 3 (HOMG3) (OMIM Entry –

*248250) due to a defect in renal tubular transport of magnesium, Claudins such as CLDN19 (OMIM Entry *603959) due to transmembrane proteins interacting with the function of tight junctions, or different types of Dent disease (OMIM Entries *300008, *300009, *310468) due to misfunction of voltage-gated chloride ion channel (CLC-5). Another reason of hypercalciuria is a distal renal tubular acidosis. In the autosomal dominant renal tubular acidosis syndromes (OMIM Entry *179800, *109270), hypocitric aciduria, hypokalemia, and osteomalacia are present. Other types of renal tubular acidosis are the autosomal recessive forms (OMIM Entry *602722, *605239) or those with the accompanying hearing loss (OMIM Entry *267300, *192132). In case of a suspected genetic disease, molecular genetic testing should be performed (Ammenti et al. 2006).

Secondary hypercalciuria in young infants is mainly caused by vitamin D toxicity or hyperparathyroidism. It can develop as a result of parenteral nutrition, high protein diet, or treatment with furosemide. Hypervitaminosis D, caused by excessive exogenous supplementation, can also induce hypercalcemia and hypercalciuria. However, most studies do not support the presence of a significant association between a higher serum level of 25(OH)D and increased risk of urinary stone formation. Tang and Chonchol (2013) have demonstrated that a short-term nutritional vitamin D_3 supplementation in patients with 25(OH)D deficiency does not increase urinary calcium excretion.

In the current study, we found that most of the infants with urolithiasis had hypercalcemia. The serum level of calcium was higher in boys than in girls, but it failed to associate with the serum 25 (OH)D level. We excluded the possibility of hypercalcemia due to the presence of the Williams-Beuren syndrome (OMIM Entry *194050) in any of the infants. This study also demonstrates that most of the infants with urolithiasis had the 25(OH)D content within the normal range and also normal activity of PTH. The serum content of 25(OH)D was increased in only three (9.4%) infants. Hypercalcemia and hypercalciuria in these infants could be ascribed to the overdosing of vitamin D_3. Infants with hypercalcemia and a normal level of 25(OH)D have, in all likelihood, idiopathic hypercalcemia (Nesterova et al. 2013). The potential reason for idiopathic infantile hypercalcemia is mutations in the *CYP24A1* gene that encodes for the enzyme converting the active forms of vitamin D to inactive metabolites (Ketha et al. 2015). Idiopathic hypercalcemia manifests during infancy with acute episodes of hypercalcemia, hypercalciuria, and nephrocalcinosis (Dinour et al. 2013; Gigante et al. 2016).

A limitation of the current study was a small group of patients from a single clinical center. We also encountered some difficulties in the assessment of calcium excretion. The most important tool in the diagnosis of hypercalciuria is a 24-h urine collection, which is usually difficult in infants, particularly that bladder catheterization is not recommended in this case. In the clinical setting, calcium excretion is most often assessed from random urine samples, and it may somehow vary depending on the diet, medication, and fluid intake.

Despite these limitations we believe we have demonstrated that the most important risk factor for urolithiasis in infants is hypercalcemia. Particular attention should be paid to infants with hypercalcemia who receive milk formula and additional vitamin D_3 supplementation. This study also demonstrates that hypercalciuria was a dominant, but not ubiquitous, metabolic abnormality in urine of infants with urolithiasis. The incidence of hypercalciuria was lower than that

hypercalcemia. Interestingly, majority of infants with urolithiasis had a normal serum level of 25 (OH)D. A low incidence of hypercalciuria detected in infants could be secondary to stone formation in the urinary tract. We believe that the possibility of *CYP24A1* gene mutations should be taken into consideration in the diagnostics of infants with urolithiasis, which may have relevance to foreseeing the risk of urolithiasis in adulthood.

Conflicts of Interest The authors declare that they have no conflicts of interest in relation to this article.

Ethical Approval All procedures performed in studies involving human participants were in accordance with the ethical standards of the institutional and/or national research committee and with the 1964 Helsinki declaration and its later amendments or comparable ethical standards.

Informed Consent Informed consent was obtained from all parents/guardians of all the individual participants included in the study.

References

Ammenti A, Neri E, Agistri R, Beseghi U, Bacchini E (2006) Idiopathic hypercalciuria in infants with renal stones. Pediatr Nephrol 21(12):1901–1903

Dinour D, Beckerman P, Ganon L, Tordjaman K, Eisenstein Z, Holtzman EJ (2013) Loss-of–function mutations of CYP24A1, the vitamin D 24–hydroxylase gene, cause long–standing hypercalciuric nephrolitiasis and nephrocalcinosis. J Urol 190 (2):552–557

Ergon EY, Akil İO, Taneli F, Oran A, Ozyurt BC (2017) Etiologic risk factors and vitamin D receptor gene polymorphisms in under one–year–old infants with urolithiasis. Urolithiasis 46(4):349–356

Fallahzadeh MH, Zare J, GH A–H, Derakhshan A, Basiratnia M, Arasteh MM, Fallahzadeh MA, Fallahzadeh MK (2012) Elevated serum levels of Vitamin D in infants with urolithiasis. Iran J Kidney Dis 6 (3):186–191

Gigante M, Santangelo L, Diella S, Caridi G, Argentiero L, D'Alessandro MM, Martino M, Stea ED, Ardissino G, Carbone V, Pepe S, Scrutinio D, Maringhini S, Ghiggeri GM, Grandaliano G, Giordano M, Gesualdo L (2016) Mutational spectrum of CYP21A1 gene in a cohort of Italian patients with idiopathic infantile hypercalcemia. Nephron 133 (3):193–204

Habbig S, Beck BB, Hoppe B (2011) Nephrocalcinosis and urolithiasis in infants. Kidney Int 80 (12):1278–1291

Hesse A (2005) Reliable data from diverse regions of the world exist to show that there has been a steady increase in the prevalence of urolithiasis. World J Urol 23:302–303

Holick MF, Binkley NC, Bischoff–Ferrari HA, Gordon CM, Hanley DA, Heaney RP, Murad MH, Weaver CM, Endocrine Society (2011) Evaluation, treatment, and prevention of vitamin D deficiency: an Endocrine Society clinical practice guideline. J Clin Endocrinol Metab 96:1911–1930

Hoppe B, Kemper MJ (2010) Diagnostic examination of the child with urolithiasis or nephrocalcinosis. Pediatr Nephrol 25:403–413

IOM (2011) Dietary reference intakes for calcium and vitamin D. Institute of Medicine of the US National Academy of Sciences. The National Academic Press, Washington, DC

Jobs K, Straż-Żebrowska E, Placzyńska M, Zdanowski R, Kalicki B, Lewicki S, Jung A (2014) Interleukin–18 and NGAL in assessment of ESWL treatment safety in infants with urolithiasis. Cent Eur J Immunol 39 (3):384–391

Ketha H, Wadams H, Lteif A, Singh RJ (2015) Iatrogenic vitamin D toxicity in an infant – a case report and review of literature. J Steroid Biochem Mol Biol 148:14–18

López M, Hoppe B (2010) History, epidemiology and regional diversities of urolithiasis. Pediatr Nephrol 25 (1):49–59

Marzuillo P, Guarino S, Apicella A, La Manna A, Polito C (2017) Why we need a higher suspicion index of urolithiasis in infants. J Pediatr Urol 13(2):164–171

Milliner DS, Murphy ME (1993) Urolithiasis in pediatric patients. Mayo Clin Proc 68:241–248

Nesterova G, Malicdan MC, Yasuda K, Sakaki T, Vilboux T, Ciccone C, Horst R, Huang Y, Golas G, Introne W, Huizing M, Adams D, Boerkoel CF, Collins MT, Gahl WA (2013) 1,25–(OH)2D–24 hydroxylase (CYP24A1) deficiency as a cause of nephrolithiasis. Clin J Am Soc Nephrol 8(4):649–657

Pludowski P, Holick MF, Grant WB, Konstantynowicz J, Mascarenhas MR, Haq A, Povoroznyuk V, Balatska N, Barbosa AP, Karonova T, Rudenka E, Misiorowski W, Zakharova I, Rudenka A, Łukaszkiewicz J, Marcinowska–Suchowierska E, Łaszcz N, Abramowicz P, Bhattoa HP, Wimalawansa SJ (2018) Vitamin D supplementation guidelines. J Steroid Biochem Mol Biol 175:125–135

Skalova S, Lutilek S (2006) High urinary N–Acetyl–beta–D–glokosaminidase activity and normal calciuria in infants with nocturnal enuresis. Indian Pediatr 43:655–656

Tang J, Chonchol MB (2013) Vitamin D and kidney stone disease. Curr Opin Nephrol Hypertens 22(4):383–389

Unal AD, Tarcin O, Parildar H, Cigerli O, Eroglu H, Nilgun Guvener Demirag NG (2014) Vitamin D deficiency is related to thyroid antibodies in autoimmune thyroiditis. Cent Eur J Immunol 39(4):493–497

Van't Hoff WG (2004) Aetiological factors in paediatric urolithiasis. Nephron Ciln Pract 98:c45–c48

Walther PC, Lamm D, Kaplan GW (1980) Pediatric urolithiasis: a ten–year review. Pediatrics 65:1068–1072

Weigert A, Hoppe B (2018) Nephrolithiasis and nephrocalcinosis in childhood–risk factor–related current and future treatment options. Front Pediatr 6:98

Zafar MN, Ayub S, Tanwri H, Naqvi SAA, Rizvi SAH (2018) Composition of urinary calculi in infants: a report from an endemic country. Urolithiasis 46 (5):445–452

Adv Exp Med Biol - Clinical and Experimental Biomedicine (2019) 4: 83–96
https://doi.org/10.1007/5584_2018_324
© Springer Nature Switzerland AG 2019
Published online: 26 January 2019

Morphometric Analysis of the Lumbar Vertebrae Concerning the Optimal Screw Selection for Transpedicular Stabilization

Jarosław Dzierżanowski, Monika Skotarczyk,
Zuzanna Baczkowska-Waliszewska, Mateusz Krakowiak,
Marek Radkowski, Piotr Łuczkiewicz, Piotr Czapiewski,
Tomasz Szmuda, Paweł Słoniewski, Edyta Szurowska,
Paweł J. Winklewski, Urszula Demkow,
and Arkadiusz Szarmach

Abstract

Transpedicular stabilization is a frequently used spinal surgery for fractures, degenerative changes, or neoplastic processes. Improper screw fixation may cause substantial vascular or neurological complications. This study seeks to define detailed morphometric measurements of the pedicle (height, width, and surface area) in the aspects of screw length and girth selection and the trajectory of its implantation, i.e., sagittal and transverse angle of placement. The study was based on CT examinations of 100 Caucasian patients (51 women and 49 men) aged 27–75 with no anatomical, degenerative, or post-traumatic spine changes. The results were stratified by gender and body side, and they were counter compared with the available literature

J. Dzierżanowski, M. Krakowiak, T. Szmuda,
and P. Słoniewski
Department of Neurosurgery, Faculty of Medicine,
Medical University of Gdansk, Gdansk, Poland

M. Skotarczyk, Z. Baczkowska-Waliszewska,
E. Szurowska, and A. Szarmach
Second Department of Radiology, Faculty of Health
Sciences, Medical University of Gdansk, Gdansk, Poland

M. Radkowski
Department of Immunopathology of Infectious and
Parasitic Diseases, Medical University of Warsaw,
Warsaw, Poland

P. Łuczkiewicz
Second Clinic of Orthopaedics and Kinetic Organ
Traumatology, Faculty of Medicine, Medical University of
Gdansk, Gdansk, Poland

P. Czapiewski
Department of Pathomorphology, Faculty of Medicine,
Medical University of Gdansk, Gdansk, Poland

Department of Pathology, Otto-von-Guericke University,
Magdeburg, Germany

P. J. Winklewski
Second Department of Radiology, Faculty of Health
Sciences, Medical University of Gdansk, Gdansk, Poland

Department of Human Physiology, Faculty of Health
Sciences, Medical University of Gdansk, Gdansk, Poland

Department of Clinical Anatomy and Physiology, Faculty
of Health Sciences, Pomeranian University in Slupsk,
Slupsk, Poland

U. Demkow (✉)
Department of Laboratory Diagnostics and Clinical
Immunology of Developmental Age, Medical University
of Warsaw, Warsaw, Poland
e-mail: demkow@litewska.edu.pl

database. Pedicle height decreased from L1 to L4, ranging from 15.9 to 13.3 mm. Pedicle width increased from L1 to L5, extending from 6.1 to 13.2 mm. Pedicle surface area increased from L1 to L5, ranging from 63 to 140 mm^2. Distance from the point of entry into the pedicle to the anterior surface of the vertebral body, defining the maximum length of a transpedicular screw, varied from 54.0 to 50.2 mm. Variations concerning body sides were inappreciable. A transverse angle of screw trajectory extended from 20° to 32°, shifting caudally from L1 to L5, with statistical differences in the L3–L5 segments. A sagittal angle varied from 10° to 12°, without such definite relations. We conclude that the L1 and L2 segments display the most distinct morphometric similarities, while the greatest differences, in both genders, are noted for L3, L4, and L5. The findings enable the recommendation of the following screw diameters: 4 mm for L1–L2, 5 mm for L3, 6 mm for L4–L5, and the length of 50 mm. We believe the study has extended clinical knowledge on lumbar spine morphometry, essential in the training physicians engaged in transpedicular stabilization.

Keywords

Bone screw fixation · Computed tomography · Lumbar vertebrae · Pedicle · Spinal surgery · Transpedicular stabilization

1   Introduction

In annual medical records, there is an increasing number of cases with spinal injuries due to traffic or work-related accidents, society aging, and multifarious age-related diseases that require neurosurgical interventions. Treatment mainly consists of transpedicular stabilization. Posterior approach for the vertebral stabilization is done by inserting screws through pedicles into the vertebral body and fixation by titan rods. That enables the correction of the spine curvature and stabilization of the vertebral level. Intraoperative radiological

neuronavigation such as fluoroscopy, O-arm, or computed tomography (CT), and electrophysiological neural monitoring reduces the risk of inappropriate positioning of the screws and related complications (Tannoury et al. 2005). However, availability of such methods is generally limited to large traumatic centers, and surgery has often been done in most hospitals with orthopedic or neurosurgical departments. A missinsertion of screws may be associated with damage to the nerve roots, dural sack (fluid), spinal cord, intercostal arteries, pleura, pedicle fracture, and others (Kleck et al. 2016; Phan et al. 2015; Villavicencio et al. 2015; Yee et al. 2008; Sponseller et al. 2008). In many cases, damage may be caused by a selection of an improper length or diameter of screws. A morphometric knowledge of the pedicle, angle of entry into the pedicle, trajectory of a screw passage through the bone vertebral elements, potential safety zone, and the awareness of individual anatomical variability of lumbar sections are essential for avoiding intraoperative complications (Kunkel et al. 2010).

This study seeks to define the anatomical elements of the vertebrae and to recommend the screw size to stabilize the posterior L1–L5 spine segments in a population of adult Caucasians, taking into account gender and body side. The results were compared with those known for other human races, based on the available literature.

2   Methods

2.1   Patients and Radiological Imaging Parameters

The study was based on CT examinations of 100 patients (51 women and 49 men) aged 27–75 (mean age 35 years) hospitalized in the Department of Neurosurgery at the Medical University in Gdansk, Poland, in January–April 2016. The reason for hospitalization was the diagnostics of the underlying cause of protracted spinal pain complaints. The choice of CT was based on the high sensitivity, noninvasiveness, and reproducibility of morphometric measurements of this radiologic technique. From this group, patients with diagnosed degenerative disease of the spine (spondylopathy, hypertrophy of the joints and ligaments), scoliosis, diseases destroying the structure of bone tissue (tumors, osteoporosis), and fractures, both posttraumatic and pathological, were excluded. Likewise, patients with a history of disc herniation, vertebral fractures, congenital or acquired bone deformities, or tumors were excluded. Only patients with a normal radiological picture and normal body mass index (BMI) were selected for the study, which created a theoretical group of "healthy" spines. The measurements concerned six anatomical elements, based on a study of Wolf et al. (2001), which was later reflected in the selection of surgical instruments and techniques. Using the bone window on scans, the following parameters were measured: the axial plane – 1/ pedicle width (PW) in the isthmus; 2/ transverse angle α, contained between the median line passing through the center of the spinous process and transpedicular screw passing through the center of the pedicle; 3/ distance (D) from the point of entry into the pedicle to the anterior surface of the vertebral body, defining the maximum length of the potential transpedicular screw in the sagittal plane; 4/ pedicle height (PH) at its narrowest point (isthmus); 5/ sagittal angle β,

contained between the pedicle axis projection and the transverse plane in the anatomic position and coronal plane; 6/ pedicle surface area (PA).

The exemplary measurement points in CT images are depicted on Figs. 1 and 2. The measurements were performed bilaterally for the L1–L5 vertebrae, which gave a total of 1000 results. Statistical analysis was based on the differences between the investigated lumbar spine levels (from L1 to L5), the gender, and the examined side (right or left). The reading accuracy was 0.1 mm for both linear and angular numerical values. All measurements were performed independently by two radiology specialists with many years of clinical experience.

2.2 Imaging Protocol and Image Post-Processing

Non-contrast CT images of Th11 to L5 spine were taken in 5-mm contiguous axial sections, with the following imaging parameters: 140 kVp, 335 mA, 0.9s rotation time, number of images of 56, total exposure time of 6.3 s, and the computed tomography dose index between 50 and 60 mGy. Raw datasets were loaded into a dedicated diagnostic workstation (AW 4, GE

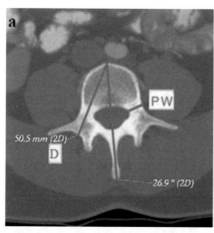
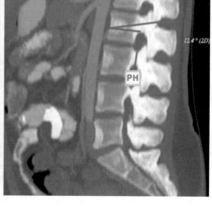

Fig. 1 Measurements performed in CT images: (**a**) axial plane, distance from the entry point of the pedicle to the anterior surface of the vertebral body (D), pedicle width (PW), and the transverse angle α of screw implantation; (**b**) sagittal plane, angle β of screw implantation and pedicle height (PH)

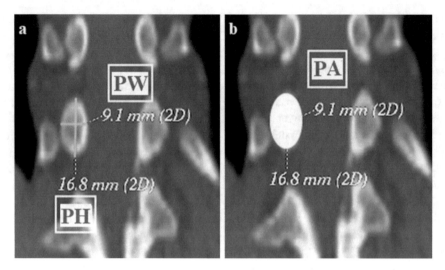

Fig. 2 Measurements performed in the coronal plane of CT images: (**a**) pedicle width (PW) and height (PH); (**b**) pedicle area (PA)

Healthcare Technologies, WI) equipped with a professional post-processing software package to generate three-dimensional images of thoracic and lumbar spine.

2.3   Statistical Analysis

Data are means ±SD and ranges. Data distribution was assessed with the Shapiro-Wilk test and equality of variances with the Levine test. Continuous variables were compared using a t-test, if normally distributed, or Mann-Whitney U test, if non-normally distributed. One-way ANOVA was used to assess differences between the means of more than two measurements between spine levels. A p-value <0.05 defined statistically significant differences. A commercial Statistica v12.0 (StatSoft, Tulsa, OK) or GraphPad Prism v6.05 (La Jolla, CA) package was used for statistical analysis.

3   Results

3.1   Lumbar Spine Measurements

The mean gender-specific results of the lumbar spine indices evaluated in this study are presented in Tables 1 and 2.

3.2   Gender-Dependent Differences in Lumbar Spine Measurements

Statistical evaluation of inter-gender differences in the lumbar spine indices stratified by the lumbar spine level is presented in Table 3 (panels a–e).

3.3   Analysis of Literature on Lumbar Spine Morphometry

The current results were subjected to a comparative analysis with literature data on lumbar spine morphometry, derived from the Internet search engine covering Medline, Embase, and PubMed. The articles taken into consideration regarded the measurements performed in radiological images, in cadavers, and finite element models based on the adult human spine. Case reports without a literature review, abstracts, intraoperative measurements assumed to be taken in sick spines, and pediatric work were all excluded from the search. The results tallied are presented in Tables 4, 5, 6, 7, and 8.

In the present study, PH decreased from L1 to L5 of the lumbar column in a range of 14.4–13.6 mm in women and 15.9–14.4 mm in men. However, there was no statistically

Table 1 Pedicle parameters measured in the lumbar bodies L1, L2, L3, L4, and L5 in women

Parameter	Side	L1	L2	L3	L4	L5
PH	R	14.5 ± 1.0	14.2 ± 1.2	13.5 ± 1.1	13.4 ± 1.1	13.4 ± 1.9
	L	14.2 ± 1.4	14.0 ± 1.1	13.6 ± 1.2	13.1 ± 1.2	13.7 ± 0.9
PW	R	5.6 ± 1.0	6.2 ± 1.1	7.5 ± 2.4	9.7 ± 1.5	13.6 ± 1.0
	L	5.6 ± 1.2	6.1 ± 1.3	7.7 ± 1.4	9.7 ± 1.1	12.8 ± 2.3
D	R	52.7 ± 2.9	52.8 ± 3.5	53.1 ± 4.0	50.9 ± 3.9	51.0 ± 4.6
	L	51.1 ± 3.7	51.9 ± 4.6	52.9 ± 4.4	51.3 ± 3.9	51.9 ± 4.2
Angle α	R	21.5 ± 2.4	21.0 ± 2.0	22.7 ± 2.8	26.3 ± 2.3	32.3 ± 3.5
	L	20.8 ± 2.3	21.0 ± 2.5	23.3 ± 2.4	25.9 ± 1.6	32.1 ± 3.6
Angle ß	R	12.6 ± 1.2	12.6 ± 1.2	11.7 ± 1.0	11.9 ± 1.3	11.2 ± 1.8
	L	12.2 ± 1.1	11.9 ± 1.4	11.5 ± 1.0	11.4 ± 1.1	10.8 ± 1.5
PA	R	63.7 ± 15.1	68.1 ± 13.9	85.4 ± 16.3	101.5 ± 15.8	140.5 ± 26.9
	L	67.2 ± 17.2	71.1 ± 13.1	84.5 ± 14.3	105.2 ± 17.3	137.4 ± 25.4

Data are means ±SD in the unit of lengths (mm) and a corresponding unit of area or in the unit of degree for angular dimensions. *PH* pedicle height; *PW* pedicle width; *D* distance from the entry point into the pedicle to the anterior surface of the vertebral body; *PA* pedicle area; *α* transverse angle of the pedicle angulation, i.e., the angle between trajectory of the transpedicle screw and the midline of vertebra; *β* sagittal angulation, i.e., the angle between trajectory of the transpedicle screw and the anatomical transverse plane of the spine; *M* male, *F* female, *R* right side, *L* left side

Table 2 Pedicle parameters measured in the lumbar bodies L1, L2, L3, L4, and L5 in men

Parameter	Side	L1	L2	L3	L4	L5
PH	R	16.2 ± 1.0	15.4 ± 1.3	15.0 ± 1.0	14.5 ± 1.3	14.3 ± 1.2
	L	15.6 ± 1.0	15.3 ± 1.2	14.8 ± 1.3	14.1 ± 1.6	14.5 ± 1.1
PW	R	6.7 ± 1.9	7.4 ± 2.4	9.0 ± 2.6	10.5 ± 2.6	13.5 ± 2.8
	L	6.7 ± 1.9	7.1 ± 2.1	8.8 ± 2.5	10.4 ± 2.4	13.2 ± 2.8
D	R	54.0 ± 3.6	54.5 ± 3.6	53.6 ± 4.0	51.2 ± 3.4	50.9 ± 4.4
	L	53.8 ± 4.1	53.3 ± 3.9	53.3 ± 4.6	51.5 ± 3.8	51.0 ± 5.2
Angle α	R	12.3 ± 1.4	12.2 ± 1.0	11.7 ± 1.0	12.1 ± 1.2	11.45 ± 1.6
	L	12.1 ± 1.2	19.9 ± 2.8	11.9 ± 1.0	11.5 ± 1.7	11.2 ± 1.7
Angle ß	R	20.3 ± 3.1	20.6 ± 2.2	22.6 ± 2.8	26.2 ± 3.8	32.15 ± 5.1
	L	19.5 ± 3.3	19.9 ± 2.8	22.5 ± 3.3	26.1 ± 3.6	31.83 ± 4.8
PA	R	83.6 ± 24.7	87.7 ± 30.0	104.3 ± 32.1	117.1 ± 31.7	138.9 ± 47.2
	L	81.0 ± 20.1	91.4 ± 26.3	101.3 ± 29.9	120.1 ± 36.6	142.1 ± 42.1

Data are means ±SD in the unit of lengths (mm) and a corresponding unit of area or in the unit of degree for angular dimensions. *PH* pedicle height; *PW* pedicle width; *D* distance from the entry point into the pedicle to the anterior surface of the vertebral body; *PA* pedicle area; *α* transverse angle of the pedicle angulation, i.e., the angle between trajectory of the transpedicle screw and the midline of vertebra; *β* sagittal angulation, i.e., the angle between trajectory of the transpedicle screw and the anatomical transverse plane of the spine; *M* male, *F* female, *R* right side, *L* left side

significant gender difference in pedicle height concerning L1 and L2, as opposed to L3 to L5 where the decreases significantly differed between women and men (Table 3). A progressive decrease in pedicle height has been also noticed by others authors (Acharya et al. 2010; Kadioglu et al. 2003; Zindrick et al. 1987; Ebraheim et al. 1996; Olsewski et al. 1990). Yet Mahato (2011), Lien et al. (2007), Noijri et al. (2005), Hou et al. (1993), and Panjabi et al. (1992) have failed to observe such a decrease, using various measurement techniques such as magnetic resonance imaging (MRI), CT, or the assessment in cadavers (Table 6).

In contradistinction we found that PW increased in both gender groups from L1 to L5 of the lumbar column in a range of 6.7–13.3 mm in women and 5.6–13.1 mm in men. Akin to pedicle height, we found no significant gender difference in pedicle width increases concerning

Table 3 Analysis of the parameters tested in the current study stratified by the lumbar spine level L1 to L5 and by gender

Parameter	Male	n	Female	n	p <
a					
PHa	15.9 ± 1.0	98	14.4 ± 1.2	102	0.00001
PW	6.7 ± 1.9	98	5.6 ± 1.1	102	0.00001
D	53.9 ± 3.9	94	51.9 ± 3.4	102	0.00014
Angle α	19.9 ± 3.2	96	21.2 ± 2.3	88	0.00214
Angle β	12.2 ± 1.3	92	12.4 ± 1.2	102	0.26005
PA	83.6 ± 24.7	98	63.7 ± 15.1	102	0.00001
b					
PH	15.3 ± 1.3	98	14.1 ± 1.2	102	0.00001
PW	7.2 ± 2.3	98	6.1 ± 1.2	102	0.00003
D	53.9 ± 3.8	94	52.3 ± 4.1	102	0.00500
Angle α	20.2 ± 2.5	96	21.0 ± 2.3	88	0.02690
Angle β	12.0 ± 1.1	92	12.3 ± 1.3	102	0.18500
PA	87.7 ± 30.0	98	68.1 ± 14.0	102	0.00001
c					
PH	14.9 ± 1.2	98	13.8 ± 1.1	102	0.00001
PW	8.9 ± 2.6	98	7. 9 ± 1.4	102	0.00058
D	53.4 ± 4.3	94	52.3 ± 4.1	102	0.07450
Angle α	22.5 ± 3.1	96	23.1 ± 2.1	88	0.12376
Angle β	11.8 ± 1.0	92	11.9 ± 1.2	102	0.69281
PA	104.3 ± 32.1	98	85.4 ± 16.3	102	0.00001
d					
PH	14.3 ± 1.5	98	13.3 ± 1.2	102	0.00001
PW	10.4 ± 2.5	98	9.7 ± 1.4	102	0.01337
D	51.4 ± 3.6	94	51.1 ± 3.9	102	0.62706
Angle α	26.1 ± 3.7	96	26.1 ± 2.0	88	0.92968
Angle β	11.8 ± 1.3	92	11.7 ± 1.2	102	0.54058
PA	117.1 ± 31.7	98	101.5 ± 15.8	102	0.00002
e					
PH	14.4 ± 1.1	98	13.6 ± 1.0	102	0.00001
PW	13.3 ± 2.8	98	13.1 ± 2.2	102	0.52219
D	51.0 ± 4.8	94	51.4 ± 4.4	102	0.50926
Angle α	32.0 ± 4.9	96	32.2 ± 3.5	88	0.77931
Angle β	11.3 ± 1.6	92	11.0 ± 1.6	102	0.11194
PA	138.8 ± 47.2	98	140.5 ± 26.9	102	0.76319

Data are means ±SD in the unit of lengths (mm) and a corresponding unit of area or in the unit of degree for angular dimensions. *PH* pedicle height; *PW* pedicle width; *D* distance from the entry point into the pedicle to the anterior surface of the vertebral body; *AP* pedicle area; *α* transverse angle of the pedicle angulation, i.e., the angle between trajectory of the transpedicle screw and the midline of vertebra; *β* sagittal angulation, i.e., the angle between trajectory of the transpedicle screw and the anatomical transverse plane of the spine, Mann-Whitney U test was used for the comparison of group means in all of the panels

L1 and L2 either, although such differences were present for L3 to L5 values (Table 3). Decreases in pedicle width we noticed unequivocally conform with other scientific articles reporting on the subject (Alam et al. 2014; Acharya et al. 2010; Abuzayed et al. 2010; Lien et al. 2007; Kadioglu et al. 2003; Chadha et al. 2003; Panjabi et al. 1992; Mitra et al. 2002; McLain et al. 2001; Mughir et al. 2010; Ebraheim et al. 1996; Kim et al. 1994; Cheung et al. 1994; Hou et al. 1993; Olsewski et al. 1990; Zindrick et al. 1987; Berry et al. 1987 (Table 7). A comparative analysis of

Table 4 Literature-based analysis of transpedicular screw angle fixation in the lumbar vertebral bodies L1–L5 in the transverse projection. All parameters were measured in degrees

Author	Methodology	α L1	α L2	α L3	α L4	α L5
Oertel et al. (2011)	94 screws transpedicular stabilization	17.0 ± 7.4	20.0 ± 4.3	19.5 ± 4.8	17.1 ± 5.0	17.6 ± 4.6
Abuzayed et al. (2010)	48 CT scans	13.1 ± 5.3	14 ± 5.2	18.0 ± 4.4	22 ± 5.4	33.3 ± 10.4
Ebraheim et al. (1996)	50 lumbar specimens	25.8 ± 2.2 M	27.3 ± 2.3 M	29.4 ± 2.3 M	33.6 ± 2.3 M	40.6 ± 2.6 M
		24.9 ± 2.6 F	27.0 ± 2.9 F	29.1 ± 2.5 F	33.0 ± 3.2 F	39.6 ± 3.2 F
Mughir et al. (2010)	74 CT scans	14.5 M	15.7 M	20.4 M	26.3 M	33.6 M
		13.1 F	14.6 F	19.9 F	25.6 F	33.6 F
Acharya et al. (2010) (Indians)	50 CT scans	10.9 ± 3.1	12.12 ± 2.9	15.4 ± 3.6	18.4 ± 4.3	24.7 ± 3.8
Alam et al. (2014) (Pakistani)	49 CT scans	13.1 MR; 14.1 FR	13.9 MR; 13.9 FR	16.2 MR; 17.6 FR	16.2 MR; 17.6 FR	22.5 MR; 20.1 FR
		13.2 ML;'14.8 FL	13.9 ML; 14.4 FL	16.8 ML 17.4 FL	16.8 ML 17.4 FL	23.1 ML 21.8 FL
Chadha et al. (2003)	14–20 CT scans	8.8 ± 5.8	10.0 ± 4.3	12.3 ± 4.9	15.4 ± 5.7	24.3 ± 7.0
Lien et al. (2007) (Asians)	15 spine columns	8.3 ± 3.1 L	11.5 ± 2.7 L	14.1 ± 3.4 L	19.7 ± 3.2 L	25.3 ± 4.7 L
		8.6 ± 2.5 R	13.2 ± 2.8 R	16.4 ± 3.7 R	18.1 ± 3.4 R	23.4 ± 3.9 R
Cheung et al. (1994) (Asians)	100 CT scans	16.0 ± 2.9	15.9 ± 5.0	19.2 ± 3.6	22.8 ± 4.4	28.5 ± 3.9
Kadioglu et al. (2003)	29 CT scans	13.0 ± 0.7	15.1 ± 1.5	13.4 ± 4.4	15.3 ± 3.1	16.8 ± 2.5
	16 cadaveric specimens	9.0 ± 2.8	11.3 ± 2.7	12.2 ± 4.4	11.2 ± 3.9	12.6 ± 3.7
Zindrick et al. (1987)	26–76 CT scans	10.9 ± 2.2	12.0 ± 3.5	14.4 ± 3.8	17.7 ± 5.2	29.8 ± 6.3
Panjabi et al. (1992)	60 vertebrae	2.9 ± 0.7 L	2.1 ± 0.6 L	2.4 ± 0.7 L	3.0 ± 1.2 L	5.7 ± 1.5 L
		2.2 ± 0.7 R	3.3 ± 0.7 R	2.9 ± 1.1 R	4.8 ± 1.1 R	5.2 ± 1.8 R
Mitra et al. (2002)	20 cadaveric specimens	9.0 M; 5.5 F	10.1 M; 10.0 F	12.3 M; 10.5 F	14.7 M; 8.5 F	29.3 M; 21 F
Olsewski et al. (1990)	29–45 cadaveric specimens	7 ± 2 M; 5 ± 2 F	7 ± 2 M; 6 ± 2 F	8 ± 2 M; 7 ± 2 F	11 ± 4 M; 10 ± 4 F	17 ± 7 M; 18 ± 8 F
	41–42 CT scans	6 ± 2 M; 6 ± 2 F	6 ± 2 M; 8 ± 4 F	7 ± 3 M; 13 ± 4 F	11 ± 5 M; 16 ± 5 F	22 ± 5 M; 29 ± 6 F
Wolf et al. (2001) (Israel)	55 CT scans	11.8 ± 1.3	11.0 ± 1.7	12.8 ± 2.2	14.1 ± 2.1	18.5 ± 3.9

Data are means ±SD. α pedicle angulation, i.e., the angle between trajectory of the transpedicle screw and the midline of vertebra, *M* male, *F* female, *R* right side, *L* left side

the literature on the Caucasian population demonstrates that most morphometric lumbar similarities concern the L1 and L2 vertebral bodies, whereas the pedicle structure varies the most in the caudal part of lumbar spine, irrespective of the race.

We further found that the distance from the entry point into the pedicle to the anterior surface of the vertebral body, otherwise described as the maximum length of a screw, varies. It measures 50 mm in total, with a range of 51.0–53.9 mm. Analogous results have been

Table 5 Literature-based analysis of transpedicular screw angle fixation in the lumbar vertebral bodies L1–L5 in the sagittal projection. All parameters were measured in degrees

Author	Methodology	β L1	β L2	β L3	β L4	β L5
Abuzayed et al. 2010*	48 CT scans	19.4 ± 8.8	16.4 ± 7.2	15.6 ± 10.1	12.6 ± 4.5	12.1 ± 5.7
Ebraheim et al. 1996**	50 lumbar specimens	6.7 ± 1.0 M	5.1 ± 0.9 M	3.9 ± 0.8 M	3.6 ± 0.8 M	2.7 ± 1.1 M
		6.1 ± 0.9 F	4.6 ± 0.9 F	3.6 ± 0.8 F	3.2 ± 0.9 F	2.6 ± 0.9 F
Mughir et al. 2010	74 CT scans	19.6 M; 17.8 F	18.7 M; 18.3 F	17.6 M; 17.8 F	16.8 M; 16.7 F	16.5 M; 14.9 F
Alam et al. 2014 (Pakistani)	49 CT scans	3.7 MR; 4.4 FR	4.0 MR; 4.6 FR	4.7 MR; 4.9 FR	4.7 MR; 4.9 FR	4.1 MR; 4.2 FR
		3.8 ML; 4.0 FL	4.2 ML; 4.3 FL	4.5 ML; 4.8 FL	4.5 ML; 4.8 FL	3.8 ML; 4.8 FL
Lien et al. 2007 (Asians)	15 spine columns in cadavers	5.4 ± 1.2 L	5.1 ± 0.9 L	4.6 ± 0.7 L	3.5 ± 0.8 L	2.8 ± 0.7 L
		6.8 ± 2.1 R	5.8 ± 1.5 R	5.4 ± 0.8 R	3.2 ± 0.6 R	3.0 ± 0.8 R
Kadioglu et al. 2003	29 CT scans	14.0 ± 2.6	13.8 ± 2.7	13.7 ± 2.8	14.2 ± 3.1	14.9 ± 3.2
	16 cadaveric specimens	7.2 ± 1.0	8.7 ± 2.7	8.2 ± 2.6	9.1 ± 3.3	10.0 ± 4.6
Zindrick et al. 1987	22–52 CT scans	2.4 ± 6.3	1.8 ± 5.5	0.2 ± 4.7	0.2 ± 3.9	−1.8 ± 3.5
Panjabi et al. 1992	60 vertebrae	12.4 ± 1.9 L	11.2 ± 2.0 L	17.1 ± 1.6 L	14.7 ± 2.2 L	23.2 ± 2.5 L
		16.5 ± 5.0 R	17.1 ± 3.7 R	19.8 ± 2.3 R	18.4 ± 1.7 R	25.9 ± 1.7 R
Olsewski et al. 1990	26–44 cadaveric specimens	5 ± 1 M; 6 ± 2 F	6 ± 2 M; 5 ± 2 F	6 ± 2 M; 6 ± 2 F	6 ± 2 M; 7 ± 2 F	5 ± 2 M; 8 ± 3 F
	30–42 CT scans	6 ± 2 M; 7 ± 3 F	6 ± 2 M; 6 ± 2 F	6 ± 2 M; 7 ± 3 F	7 ± 2 M; 8 ± 3 F	7 ± 4 M; 7 ± 4 F

Data are means ±SD. β sagittal angulation, i.e., the angle between trajectory of the transpedicle screw and the body of vertebra in the sagittal plane, *M* male, *F* female, *R* right side, *L* left side; *angle between the pedicle axis and the superior vertebral body border in the sagittal plane; **angle between the pedicle axis projection and the transverse plane in the anatomic position

obtained by other authors (Acharya et al. 2010; Mughir et al. 2010; Noijri et al. 2005; Chadha et al. 2003; Kadioglu et al. 2003; Hou et al. 1993; Zindrick et al. 1987) (Table 8). The PA, in its narrowest section, increased from L1 to L5, with a range of 63.7–140.5 mm² in both genders; the increases were morphometrically akin to each other down along the lumbar bodies (Tables 1, 2, and 3). The PA measurements could not be counter compared due to the lack of relevant literature data.

As the curvature of the spine column changes, the angle of screw insertion also changes. We used two angles to describe the screw location in the sagittal and transvers projections. The sagittal angle β decreased along L1 to L5 from 12.4° to 11.0°, with no appreciable difference between body sides or between women and men. Similar decreases have been noticed by other authors (Abuzayed et al. 2010; Mughir et al. 2010; Lien et al. 2007; Ebraheim et al. 1996; Panjabi et al. 1992). However, other authors report the lack of angle β decrease or even a reverse tendency (Alam et al. 2014; Kadioglu et al. 2003; Olsewski et al. 1990) (Table 5). The transverse angle α of transpedicular screw fixation increased along L1 to L5 from 19.9° to 32.2° on both left and right body side and in both genders. A similar finding has been described by some (Alam et al. 2014; Abuzayed et al. 2010; Acharya et al. 2010; Mughir et al. 2010; Lien et al. 2007; Chadha et al. 2003; Cheung et al. 1994; Zindrick et al. 1987), but not by the other authors (Oertel et al. 2011; Kadioglu et al. 2003; Panjabi et al. 1992) (Table 4).

Overall, in the current study, we show that the most significant differences in PH, PW, and PA occurred in the L3-L5 vertebral bodies, which has not been acknowledged in the available literature database reviewed herein. In the upper lumbar

Table 6 Literature-based analysis of pedicle height measured bilaterally in the lumbar spine L1–L5. All parameters were measured in millimeters

Author	Methodology	PH L1	PH L2	PH L3	PH L4	PH L5
Mahato (2011)	50 X-ray	15.8 ± 2.2 R	16.2 ± 1.9 R	16.3 ± 1.6 R	15.7 ± 1.7 R	14.2 ± 2.4 R
		15.4 ± 2.0 L	16.8 ± 1.8 L	15.9 ± 1.8 L	15.3 ± 1.9 L	15.1 ± 2.1 L
Ebraheim et al. (1996)	50 lumbar specimens	14.1 ± 1.5 M	14.2 ± 1.3 M	13.9 ± 1.4 M	12.7 ± 1.7 M	11.4 ± 1.3 M
		14.0 ± 1.0 F	13.8 ± 1.0 F	13.8 ± 1.1 F	12.8 ± 1.6 F	11.4 ± 1.5 F
Noijri et al. (2005) (Japanese)	103 cadaveric specimens	15.9 ± 2.8 M	14.8 ± 1.6 M	14.7 ± 1.3 M	15.5 ± 2.0 M	20.7 ± 3.0 M
		15.2 ± 1.4 F	14.4 ± 1.2 F	14.2 ± 1.1 F	15.0 ± 1.8 F	20.2 ± 2.3 F
Alam et al. (2014) (Pakistani)	49 CT scans	13.5 MR; 12.8 FR	13.4 MR; 12.3 FR	12.0 MR; 11.7 FR	12.0 MR; 11.7 FR	11.5 MR; 10.9 FR
		13.2 ML; 12.6 FL	13.5 ML; 11.9 FL	12.4 ML; 11.4 FL	12.4 ML; 11.4 FL	10.3 ML; 10.8 FL
Hou et al. (1993)	40 spinal columns	15.9 ± 1.4	15.4 ± 1.6	15.3 ± 1.7	15.3 ± 1.9	20.5 ± 3.6
Lien et al. (2007) (Asians)	15 spinal columns	13.6 ± 1.4 L	14.0 ± 1.7 L	13.9 ± 1.6 L	12.5 ± 2.2 L	12.3 ± 2.3 L
		13.7 ± 1.5 R	14.1 ± 1.8 R	13.9 ± 1.7 R	13.0 ± 2.3 R	12.7 ± 2.1 R
Kadioglu et al. (2003)	29 CT scans	14.7 ± 1.7	14.5 ± 2.4	13.6 ± 1.6	13.6 ± 1.8	13.4 ± 1.7
	16 cadaveric specimens	14.2 ± 1.3	14.2 ± 2.9	13.1 ± 2.4	13.0 ± 2.1	13.2 ± 1.6
Zindrick et al. (1987)	21–52 CT scans	15.4 ± 2.8	15.0 ± 1.5	14.9 ± 2.4	14.8 ± 2.1	14.0 ± 2.3
Panjabi et al. (1992)	60 vertebrae	18.8 ± 0.7 L	14.9 ± 0.5 L	14.6 ± 0.6 L	14.7 ± 0.5 L	19.2 ± 1.0 L
		15.9 ± 0.8 R	15.0 ± 0.5 R	14.2 ± 0.6 R	13.4 ± 0.2 R	18.0 ± 1.0 R
Mitra et al. (2002)	20 cadaveric specimens	15.6 M; 15.7 F	15.2 M; 15.7 F	15.0 M; 15.4 F	14.8 M; 14.7 F	15.7 M; 17.0 F
Olsewski et al. (1990)	31–47 cadaveric specimens	17.0 ± 1.7 M	16.0 ± 2.2 M	16.1 ± 2.1 M	16.4 ± 2.5 M	17.4 ± 3.4 M
	37–42 CT scans	15.3 ± 2.0 F	15.3 ± 2.0 F	15.0 ± 1.5 F	14.9 ± 1.5 F	16.2 ± 1.9 F
		16.4 ± 1.7 M	15.4 ± 1.5 M	15.4 ± 1.6 M	15.4 ± 2.0 M	16.2 ± 2.1 M
		18.2 ± 1.7 F	17.2 ± 1.7 F	16.9 ± 1.3 F	15.6 ± 1.5 F	13.8 ± 1.9 F

Data are means ±SD. *PH* pedicle height, *M* male, *F* female, *R* right side, *L* left side

segments L1 and L2, neither were multiple nor significant differences determined. Therefore, it may be surmised that the anatomical inter-individual similarity of lumbar vertebral bodies decreases caudally.

4 Discussion

The major finding of this study was that the first and second lumbar vertebral bodies show the most morphological likeness. Significant morphometric differences appear in the caudal part of the lumbar spine from L3 to L5 in both genders. The selection of the spine points measured in the present study stemmed from the following rationale: 1/ excessively thick screw may damage the pedicle or adjacent nerves; 2/ stabilization of the resistant forces in motion cannot be placed too superficially; 3/ 50–80% of the width of the vertebral body is the recommended value; and 4/ too long a screw can pierce the anterior periosteum and damage the retroperitoneal vessels of the abdomen. Neuronavigation enables the antecedent in vivo planning, due to the possibility of 3D image reconstruction, of the spine implant selection. It also increases the accuracy of the insertion of transpedicular screws and, indisputably, safety of posterior stabilization surgery. This is especially true in case of advanced deformation that causes difficulties in the identification of radiological points of pedicle surface.

Table 7 Literature-based analysis of pedicle width measured in its narrowest section in the vertebral bodies of the lumbar spine L1–L5. All parameters were measured in millimeters

Author	Methodology	PW L1	PW L2	PW L3	PW L4	PW L5
Mahato (2011)	50 X–ray	7.8 ± 1.2 R	8.2 ± 1.7 R	9.7 ± 2.0 R	12.0 ± 2.1 R	14.6 ± 2.4 R
		7.9 ± 1.3 L	8.3 ± 1.1 L	10.3 ± 1.7 L	12.7 ± 1.8 L	15.1 ± 2.4 L
Abuzayed et al. (2010)	48 CT scans	7.2 ± 1.51	8.3 ± 1.5	10.0 ± 1.5	11.8 ± 1.9	14.9 ± 2.9
Ebraheim et al. (1996)	50 lumbar specimens	7.4 ± 1.3 M	8.4 ± 1.4 M	9.8 ± 1.8 M	12.8 ± 2.0 M	18.3 ± 1.7 M
		7.5 ± 1.6 F	7.9 ± 1.2 F	9.7 ± 1.4 F	12.5 ± 1.8 F	17.6 ± 3.1 F
Mughir et al. (2010)	74 CT scans	7.3 M; 5.9 F	8.2 M; 6.4 F	10.1 M; 8.1 F	12.0 M; 10.0 F	16.2 M; 13.9 F
Noijri et al. (2005) (Japanese)	103 cadaveric specimens	7.4 ± 2.0 M	7.8 ± 1.7 M	9.1 ± 1.7 M	10.1 ± 1.7 M	11.1 ± 1.7 M
		6.9 ± 1.5 F	7.4 ± 1.5 F	8.9 ± 1.6 F	9.7 ± 1.4 F	10.6 ± 1.5 F
Acharya et al. (2010) (Indians)	50 CT scans	7.2 ± 0.9	7.6 ± 0.8	9.0 ± 1.1	11.1 ± 1.0	13.9 ± 1.2
Alam et al. (2014) (Pakistani)	49 CT scans	6.4 MR; 5.6 FR	7.3 MR; 6.4 FR	10.5 MR; 9.6 FR	10.5 MR; 9.6 FR	13.5 MR; 12.2 FR
		6.1 ML; 5.9 LF	7.3 ML; 6.4 LF	10.6 ML; 9.7 LF	10.6 ML; 9.7 LF	13.5 ML; 12.7 LF
Chadha et al. (2003)	14–20 CT scans	6.69 ± 1.55	7.26 ± 1.43	8.43 ± 1.42	10.81 ± 1.17	13.47 ± 1.43
Hou et al. (1993)	40 spinal columns	7.2 ± 1.3	7.6 ± 1.2	9.4 ± 1.6	10.8 ± 1.4	12.8 ± 2.7
Lien et al. (2007) (Asians)	15 cadaveric specimens spinal columns	6.5 ± 1.7 R	7.0 ± 1.8 R	9.0 ± 1.8 R	12.2 ± 2.3 R	17.7 ± 2.7 R
		6.4 ± 1.6 L	7.4 ± 1.7 L	9.3 ± 1.9 L	11.6 ± 2.1 L	17.5 ± 2.6 L
Cheung et al. (1994) (Asians)	100 CT scans	5.3 ± 0.6	6.7 ± 1.0	9.5 ± 1.6	11.5 ± 1.7	14.7 ± 1.9
Kadioglu et al. (2003)	29 CT scans	8.8 ± 0.4	9.7 ± 2.0	10.3 ± 2.0	10.8 ± 2.5	14.6 ± 3.8
	16 cadaveric specimens	6.4 ± 2.0	6.6 ± 2.3	8.6 ± 3.8	10.8 ± 3.3	12.4 ± 2.4
Zindrick et al. (1987)	26–56 CT scans	8.7 ± 2.3	8.9 ± 2.2	10.3 ± 2.6	12.9 ± 2.1	18.0 ± 4.1
Panjabi et al. (1992)	60 vertebrae	9.2 ± 0.9 L	8.7 ± 0.8 L	10.1 ± 0.5 L	14.7 ± 0.5 L	19.2 ± 1.0 L
		8.0 ± 0.9 R	7.8 ± 0.6 R	10.2 ± 0.7 R	13.4 ± 0.2 R	18.0 ± 1.0 R
Berry et al. (1987)	30 spinal columns	7.0 ± 1.9 L	7.4 ± 1.6 L	9.2 ± 1.3 L	10.3 ± 1.6 L	10.9 ± 3.4 L
		6.9 ± 1.7 R	7.5 ± 1.5 R	9.1 ± 1.6 R	10.4 ± 1.6 R	10.5 ± 2.9 R
Kim et al. (1994)	73 spinal columns	7.0 ± 1.5 M	7.5 ± 1.1 M	9.9 ± 1.5 M	12.7 ± 2.3 M	18.9 ± 2.5 M
		6.6 ± 1.5 F	6.9 ± 1.6 F	8.9 ± 3.0 F	11.9 ± 2.3 F	17.6 ± 2.0 F
Mitra et al. (2002)	20 cadaveric specimens	7.05 M; 5.95 F	7.85 M; 6.5 F	9.01 M; 7.55 F	11.6 M; 9.25 F	16.19 M; 12.75 F
Olsewski et al. (1990)	34–47 spinal columns	9.5 ± 2.9 M	9.6 ± 2.2 M	11.7 ± 2.5 M	14.7 ± 2.7 M	21.1 ± 3.4 M
	39–42 CT scans	7.7 ± 1.9 F	7.9 ± 1.9 F	9.6 ± 2.4 F	12.5 ± 2.3 F	18.4 ± 3.6 F
		8.2 ± 2.3 M	8.4 ± 2.1 M	10.2 ± 2.5 M	13.2 ± 2.5 M	20.1 ± 3.7 M
		8.2 ± 1.8 F	8.3 ± 1.8 F	10.0 ± 1.9 F	12.6 ± 2.3 F	16.6 ± 2.5 F

Data are means ±SD. *PW* pedicle width, *M* male, *F* female, *R* right side, *L* left side

Table 8 Literature-based analysis of axial length, i.e., distance from the posterior aspect of the laminar cortex to the anterior aspect of the vertebral body cortex along the pedicle axis of the lumbar spine L1–L5. All parameters were measured in millimeters

Author	Methodology	Axial length				
		PH L1	PH L2	PH L3	PH L4	PH L5
Noijri et al. (2005) (Japanese)	103 cadaveric specimens	42.5 ± 3.7 M	44.0 ± 3.5 M	45.0 ± 3.7 M	44.3 ± 3.6 M	43.4 ± 3.7 M
		40.6 ± 3.3 F	42.7 ± 3.4 F	43.6 ± 3.4 F	43.0 ± 3.3 F	41.2 ± 5.5 F
Mughir et al. (2010)	74 CT scans	48.6 M; 41.7 F	49.6 M; 42.9 F	49.9 M; 43.1 F	47.3 M; 40.4 F	46.2 M; 39.8 F
Acharya et al. (2010) (Indians)	50 CT scans	47.0 ± 3.4	49.0 ± 2.8	47.2 ± 3.9	47.5 ± 5.4	48.9 ± 4.4
Chadha et al. (2003)	14–20 CT scans	47.5 ± 3.3	49.1 ± 5.7	46.3 ± 6.1	46.3 ± 8.3	49.5 ± 6.2
Hou et al. (1993)	40 spinal columns	55.6 ± 3.2	55.4 ± 3.8	55.6 ± 3.5	53.9 ± 4.0	51.9 ± 3.9
Kadioglu et al. (2003)	29 CT scans	42.7 ± 2.7	42.5 ± 3.4	41.6 ± 3.1	41.3 ± 2.1	40.8 ± 3.0
	16 cadaveric specimens	42.7 ± 3.2	42.5 ± 3.1	41.6 ± 2.7	41.3 ± 3.1	40.8 ± 2.4
Zindrick et al. (1987)	26–74 CT scans	44.7 ± 4.5	45.5 ± 3.7	44.4 ± 5.1	40.7 ± 4.3	33.7 ± 5.6

Data are means ±SD. *PH* pedicle height, *M* male, *F* female

Nevertheless, even intraoperative use of modern radiological techniques such as navigation with CT, fluoroscopy, or O-arm carries a several percent risk of a screw misplacement (Klatka et al. 2017; Pastuszak et al. 2017; Aoude et al. 2015). Kleck et al. (2016) have reported a 95% frequency for the conventional technique and 85% for the neuronavigation technique of correctly inserted trans-pedicular screws. A conventional technique disadvantage can be attributed to neurological consequences such as pain, parasthesia, weakness of the senses, bladder dysfunction, and skeletal muscle weakness. Other complications include dislocation or breakage of the pedicle by the inserted screw, reported in 14–40% cases without neuronavigation, the incidence of which drops to 4.1–11.5% with the use of neuronavigation.

A problem occurs when only an intraoperative X-ray view is done or a freehand pedicle screw placement is employed. A malpositioned screw can be replaced during the surgical procedure. However, tinkering with the implanted screw involves a risk of breaking the bone structure with damage to the nerve roots or weakening the bone stability and endurance (Yee et al. 2008; Sponseller et al. 2008). These are the circumstances, sometimes arising in surgery without neuronavigation, in which the data presented in this study can be conducive to the antecedent selection of an appropriately fitting screw for a given vertebral level.

Recent years have resulted in a number of morphometric articles related to the construction of the vertebral canal, vertebral processes, facial joints, pedicular angulation, intervertebral openings, and intervertebral discs. Due to the great interest of medical industry, prosthesis manufacturers, and transpedicular stabilization systems for spinal surgery, considerable anatomical work focuses on the spine biomechanics. A limited availability of anatomical specimens and adequate anatomical laboratories, and consequently the cost of such research has led to the creation of virtual models based on finite element models (Dzierżanowski et al. 2014). The models are constructed on the basis of spine morphometry of sheep, llamas, or on the fragments of human spine. Unfortunately, the assumptions involved in the modeling show some imperfections. Firstly, some of the articles show a considerable divergence between the construction of animal and human spines. This is a result of different function specifics of an animal spine due to such aspects as the movement using four limbs, horizontal spine arrangement, reduced curvature of lordosis, and smaller intervertebral discs accompanied by more developed spine muscles, compared to those in humans. The human two limb gait pattern puts a greater pressure force on the vertebral elements, which translates into overload and degenerative diseases of the spine. McLain et al. (2001) have shown that the average angle of the base is statistically greater and the width of the pedicle is narrower in the goat, pig, dog, and sheep spine models than those in the human spine model. The differences also apply to other anatomical structures such as the vertebral bodies, canals, processes, and facial joints, ranging from over a dozen to 95%. Similar results have been obtained in a study of Mageed et al. (2013), where the difference in size of all measured spinal components is more than 40%. On the other hand, the disadvantage of models based on the human corpses is a small representativeness of this group. Such models are usually based on a very large number (hundreds) of measured points, radiologically scanned in a small number (several) of specimens. That has little correspondence to the variability of the living population, resulting from the patient's age, sex, weight, type of body buildup, or race. The limitation of the corpse method is also the fact that the tissue (spine) under study is exposed to chemicals used during tissue preparation and sample storage (ligaments, sphincters). Thus, while the obtained bone morphometry data is precise, the measurement, for example, of the proportion of the innate strength of the intervertebral disc to that exerted by the external forces and ligaments during movement may be disputed. Therefore, osteometric measurements in the models could be substituted for radiological measurements that are routinely performed during the CT or X-ray diagnostics. This idea is contained in the presented work. An additional advantage of radiographic morphometry is the possibility of having almost unlimited number of subjects matched in terms of gender, BMI, age, or race, and the possibility of an evaluation of morphometric data in relation to the features

above outlined. The pattern of radiological assessment of the lumbar spine dimensions has viable practical aspects making it recommendable in the interdisciplinary scientific fields related to biomechanics, anatomy, radiology, and surgery.

5   Conclusions

Differences in pedicle shape and dimensions from L1 to L5 vertebrae bear significance for the selection of transpedicular screws. A morphometric method of vertebral measurements herein described may serve to formulate an equation model for various types of implants, forgoing the necessity to employ substantial financial and technical resources. Taking into account the values presented, morphometric length of the transpedicular screw for L4 and L5 is equal for both genders, and for L1–L3 it should be longer for men. The recommended average value is 50 mm. The angle of entry is analogous in L3–L5 segments in both genders, but it is greater in men in L1 and L2 segments. No difference in pedicle width was noted in the vertebral body L5 in either gender. Considering the bone tissue margin, the recommended diameter of a transpedicular screw is 4 mm for L1 and L2, 5 mm for L3, and 6 mm for L5. The upper lumbar vertebrae L1 and L2 bear the greatest morphometric similarity, while the vertebrae L3, L4, and L5 present the most statistically significant diversity in either sex. We believe the study has extended clinical knowledge on lumbar spine morphometry, essential in the training physicians engaged in transpedicular stabilization.

Conflicts of Interest The authors declare that they have no conflicts of interest in relation to this article.

Ethical Approval All procedures performed in studies involving human participants were in accordance with the ethical standards of the institutional and/or national research committee and with the 1964 Helsinki declaration and its later amendments or comparable ethical standards.

Informed Consent Informed consent was obtained from all individual participants included in the study.

References

Abuzayed B, Tutunculer B, Kucukyuruk TS (2010) Anatomic basis of anterior and posterior instrumentation of the spine: morphometric study. Surg Radiol Anat 32:75–85

Acharya S, Dorje T, Sirvastava A (2010) Lower dorsal and lumbar pedicle morphometry in Indian population. Spine 35:378–384

Alam MM, Waqas M, Shallwani H, Javed G (2014) Lumbar morphometry: a study of lumbar vertebrae from a Pakistan population using computed tomography scans. Asian Spine J 8:421–426

Aoude AA, Fortin M, Figeiredo R, Jarzem P, Ouellet J, Weber MH (2015) Methods to determine pedicle screw placement accuracy in spine surgery: a systematic review. Eur Spine J 24:990–1004

Berry JL, Moran JM, Berg WS, Steffee AD (1987) A morphometric study of human lumbar and selected thoracic vertebrae. Spine 12:362–367

Chadha M, Balain B, Maini L, Dhaon BK (2003) Pedicle morphology of the lower thoracic, lumbar, and S1 vertebrae: an Indian perspective. Spine 28:744–749

Cheung K, Ruan D, Chan F, Fang D (1994) Computed tomographic osteometry of Asian lumbar pedicles. Spine 19:1495–1498

Dzierżanowski J, Szarmach A, Słoniewski P, Czapiewski P, Piskunowicz M, Bandurski T, Szmuda T (2014) The posterior communicating artery: morphometric study in 3D angio–computed tomography reconstruction. The proof of the mathematical definition of the hypoplasia. Folia Morphol (Warsz) 73:286–291

Ebraheim NA, Rollins JR, Xu R, Yeasting RA (1996) Projection of the lumbar pedicle and its morphometric analysis. Spine 21:1296–1300

Hou S, Hu R, Shi Y (1993) Pedicle morphology of the lower thoracic and lumbar spine in a Chinese population. Spine 18:1850–1855

Kadioglu H, Tacki E, Levent M, Arik M, Aydin I (2003) Measurements of the lumbar pedicles in the Eastern Anatolian population. Surg Radiol Anat 25:120–126

Kim NH, Lee HM, Chung IH, Kim HJ, Kim SJ (1994) Morphometric study of the pedicles of thoracic and lumbar vertebrae in Koreans. Spine 19:1390–1394

Klatka J, Grywalska E, Hymos A, Krasowska E, Mielnik M, Siwicka–Gieroba D, Markowicz J, Trojanowski P, Olszański W, Roliński J (2017) Subpopulations of natural killer–T–like cells before and after surgical treatment of laryngeal cancer. Cent Eur J Immunol 42(3):252–258

Kleck CJ, Cullimore I, Lafleur M, Lindley E, Rentschler ME, Burger EL, Cain CM, Patel VV (2016) A new 3–dimensional method for measuring precision in surgical navigation and methods to optimize navigation accuracy. Eur Spine J 25:1764–1774

Kunkel ME, Schmidt H, Wilke HJ (2010) Prediction equations for human thoracic and lumbar vertebral morphometry. J Anat 216:320–328

Lien SB, Liou NH, Wu SS (2007) Analysis of anatomic morphometry of the pedicles and the safe zone for lumbar spine. Eur Spine J 16:1215–1222

Mageed M, Berner D, Julke H, Hohaus C, Brehm W, Gerlach K (2013) Is sheep lumbar spine a suitable alternative model for human spinal researches? Morphometrical comparison study. Lab Anim Res 29:183–189

Mahato NK (2011) Disc spaces, vertebral dimensions, and angle values at the lumbar region: a radioanatomical perspective in spines with L5–S1 transitions. J Neurosurg Spine 15:371–379

McLain FR, Yerby SA, Moseley TA (2001) Comparative morphometry of V4 vertebrae. Spine 27:200–206

Mitra SR, Datir SP, Jadhav SO (2002) Morphometric study of the lumbar pedicle in the Indian population as related to pedicular screw fixation volume. Spine 27:453–459

Mughir AM, Yusof MI, Abdullah S, Ahmad S (2010) Morphological comparison between adolescent and adult lumbar pedicles using computerized tomography scanning. Surg Radiol Anat 32:587–592

Noijri K, Matsumoto M, Chiba K, Toyama Y (2005) Morphometric analysis of the thoracic and lumbar spine in Japanese on the use of pedicle screws. Surg Radiol Anat 27:123–128

Oertel MF, Hobart J, Stein M, Schreiber V, Scharbrodt W (2011) Clinical and methodological precision of spine navigation assisted by 3D intraoperative O–arm imaging. J Neurosurg Spine 14:532–536

Olsewski JM, Simmons EH, Kallen FC, Mendel FC, Severin CM, Berens DL (1990) Morphometry of the lumbar spine: anatomical perspectives related to transpedicular fixation. J Bone Joint Surg Am 72:541–549

Panjabi MM, Goel V, Oxland T, Takata K, Duranceau J, Krag M, Price M (1992) Human lumbar vertebrae quantitative tree–dimensional anatomy. Spine 17:299–306

Pastuszak Ż, Stępień A, Tomczykiewicz K, Piusińska–Macoch R, Kordowska J, Galbarczyk D, Świstak J (2017) Limbic encephalitis – a report of four cases. Cen Eur J Immunol 42(2):213–217

Phan K, Hogan J, Maharaj M, Mobbs R (2015) Cortical bone trajectory for lumbar pedicle screw placement: a review of published reports. Ortop Surg 7:213–221

Sponseller PD, Takenaga RK, Newton P, Boachie O, Flynn J, Letko L, Betz R, Bridwell K, Gupta M, Marks M, Bastrom T (2008) The use of traction in the treatment of severe spinal deformity. Spine 33:2305–2309

Tannoury T, Crowl AC, Battaglia TC, Chan DP, Anderson DG (2005) An anatomical study comparing standard fluoroscopy and virtual fluoroscopy for the placement of C1–2 transarticular screws. J Neurosurg Spine 2:584–588

Villavicencio AT, Serxner BJ, Mason A, Nelson EL, Rajpal S, Feas N, Burneikiene S (2015) Unilateral and bilateral pedicle screw fixation in transforaminal lumbar interbody fusion: radiographic and clinical analysis. Word Neurosurg 83:553–559

Wolf A, Shoham M, Shnider M, Roffman M (2001) Morphometric study of the human lumbar spine for operation–workspace specifications. Spine 26:2472–2477

Yee A, Adjei N, Do J, Ford M, Finkelstein J (2008) Do patient expectations of spinal surgery relate to functional outcome. Clin Orthop Relat Res 466:1154–1161

Zindrick MR, Wiltse LL, Dornik A, Widell EH, Knight GW, Patwardhan AG, Thomas JC, Rothman SL, Fields BT (1987) Analysis of the morphometric characteristics of the thoracic and lumbar pedicles. Spine 12:160–166

Adv Exp Med Biol - Clinical and Experimental Biomedicine (2019) 4: 97–104
https://doi.org/10.1007/5584_2018_308
© Springer Nature Switzerland AG 2019
Published online: 11 January 2019

Stress and Dehumanizing Behaviors of Medical Staff Toward Patients

Alicja Głębocka

Abstract

Dehumanization is defined as aggressive behaviors which offend people's dignity. This phenomenon is a serious problem in medicine as it affects interpersonal relationships between medical professionals and patients, patients' well-being, and the capability of following medical recommendations. There are a few factors determining dehumanizing behaviors: infrahumanization, perceiving patients as nonhuman beings, compassion fatigue, and stress. The main goal of this study was to examine the impact of stress on dehumanizing behaviors. A quasi-experimental survey was conducted in a group of 96 nurses. The following psychometric measures were employed in the study: scale of behavioral indicators of patient's dehumanization (SBIPD), mood adjective checklist (UMACL), interpersonal reactivity index (IRI) assessing aspects of empathy, and the Eysenck personality traits (EPQ-R) questionnaires. Comparative inter-group analysis (experimental vs. control) confirmed that stress on the side of medical professionals influenced the acceptance of dehumanization; it particularly influenced the cognitive evaluations of patent dehumanizing behaviors. These evaluations have no relationship to empathy, neuroticism, and psychoticism in the control group. However, moderate correlation occurred between the patent dehumanization and neuroticism in the experimental group. The findings lead to the conclusion that stress experienced in the work setting can have an effect on dehumanizing practices in medicine. One of the best ways to combat dehumanization in medicine is to reduce stress by improving the work conditions.

Keywords

Compassion fatigue · Dehumanization · Empathy · Infrahumanization · Medicine

1 Introduction

Dehumanization is defined as aggressive behaviors which offend people's dignity. There are many examples of these behaviors such as judging, criticizing, taunting, deriding, disdaining, stigmatizing, and discriminating. Such behaviors have a significance to affect interpersonal relationships. Dehumanization affects many social and psychological consequences. On one hand, it can result in increased aggression and violence or reduced interest, understanding, and empathy for victims' abuse. On the other hand, it can lead to fluctuation of self-esteem, social isolation, and emotional disorders. For

A. Głębocka (✉)
Faculty of Psychology and Humanities, The Andrzej Frycz Modrzewski University, Cracow, Poland
e-mail: aglebocka@afm.edu.pl

these reasons, dehumanization of persons and groups has been being researched by social psychologists for many years (Haslam and Stratemeyer 2016; Vaes et al. 2012; Leyens et al. 2007). Zimbardo (1970) has suggested that dehumanization is a defensive mechanism of protecting oneself against strong negative feelings by treating other people as objects rather than persons. Leyens et al. (2001) have revealed that dehumanization could have human and animal attributions, assignable to the out-group individuals. Two types of emotions have been described: secondary, for example, pride or shame, and primary, joy and rage. Secondary emotions, described as complex cognitive processes, are uniquely human, while the primary ones are perceived as common to humans and animals (Demoulin et al. 2004). People are disposed to assign secondary emotions to the in-group rather than out-group. The non-uniquely human characteristics are common for both groups; this bias has been named as infrahumanization to emphasize that a lower human status is ascribed to the out-group (Leyens et al. 2001). Moreover, infrahumanization has also been observed when groups were compared in terms of uniquely human (e.g., morality or rationality) and non-uniquely human (e.g., impulsiveness or instinct) traits (Capozza et al. 2013; Costello and Hodson 2010).

Infrahumanization is present at a large variety of levels: nations, ethnic groups, and stigmatized social minorities. Much research has examined factors contributing to infrahumanization and dehumanization and to their consequences. Yet few surveys have been done in the area of medicine. Meanwhile, treating patients as nonhuman individuals can be observed in medical practices. This manifests itself in various forms: using baby talk when speaking to the elders, talking about patients in their presence without their participation, asking for patients' opinion only to discredit it, blocking any patients' attempts to negotiate, forcing them to strictly adhere to hospital procedures, or routinizing the medical care. All these behaviors ignore the patients' right to decide about themselves (Głębocka 2017; Capozza et al. 2016).

Haque and Waytz (2012) have identified six major reasons of dehumanization, splitting them into nonfunctional and functional causes. The nonfunctional causes are the following: (i) deindividuating practices – denying a person "identity" – occurs when patients are perceived as dependent on others and incapable of making choices, (ii) impaired patient agency refers to treating hospitalized people as incapable of planning any intentional actions, and (iii) dissimilarity that manifests itself in perceiving patients through their very nature of being ill, i.e., as less resembling one's prototypical concept of a human being. Dissimilarity also involves the labeling patients as cases rather than humans suffering from a particular disease. In addition, asymmetries in the doctor–patient dyad contribute to dehumanization.

The functional causes of dehumanization are the following:

(i) Mechanization, i.e., treatment of patients as a mechanical system consisting of interacting parts. According to these beliefs, patients are viewed as being incapable of emotional responsiveness or interpersonal warmth. Sometimes, however, deconstructing patients into organs might be useful in the diagnostic and therapeutic procedures. As Haque and Waytz (2012) suggest, "Some minimal level of dehumanization exists in clinical contexts because mechanization benefits these tasks".

(ii) Reduced empathy, i.e., medical staff is incapable of understanding and feeling the affective state of patients. Perceiving pain experienced by others usually involves two steps: Emotional sharing and then cognitively reappraising this emotion (Decety 2011; Han et al. 2008). Some research indicates that physicians do not experience even the first step of empathy.

(iii) Moral disengagement, connected with reduced empathy, is defined as a need to minimize the guilt of inflicting pain. Current medicine can reduce pain effectively; nonetheless in some instances, the infliction of pain is inescapable.

There is evidence that healthcare professionals who are deeply involved in emotional relationships with patients bear severe psychological costs. An investigation by Vaes and Muratore (2013) has demonstrated that humanizing a patient's suffering by healthcare professionals makes them more vulnerable to the symptoms of burnout, especially when they are strongly engaged in a direct work with patients. The negative effects also consist of emotional or mood contagion and secondary traumatic stress, also known as compassion fatigue. The emotional or mood contagion involves transfer of emotional states either from patients or peer workers (Bakker et al. 2001; Pearlman and Saakvitne 1995). Compassion fatigue should not be understood as a simple feeling but rather as a complex attitude toward others. It also includes the impingement stemming from the work with traumatized individuals, with the consequences such as reduced capacity to withstand the suffering of people, physical and emotional exhaustion, or a lack of empathy (Turgoose et al. 2017). A complicated mechanism of compassion involves a reflection of the mental condition of another person, an active regard for his kindness as a human being, and an emotional response. A dynamic interactive process between individuals includes awareness, perspective taking, thoughts, emotions, distress tolerance, and a motivation to relieve suffering (de Zulueta 2013). Deeply suffering patients expect the attunement and empathy from medical staff. Regrettably, healthcare professionals often distance themselves from patients. They do not provide patients with the support expected and often steer clear off difficult issues, focusing on biomedical facts.

There are many issues that cause a corrosion of compassion among caregivers. The psychological reasons involve the existential fear of death and dying; conscious experience of doubt, uncertainty, anxiety, and guilt; but also high capacity for feeling and expressing empathy (Figley 2002). However, compassion fatigue and dehumanization in medicine also stem from the structural and organizational rules of functioning of healthcare facilities. An increase in compassion fatigue, secondary traumatic stress, and dehumanizing behaviors over time could be explained by a cumulative stress model. Frequent and strong exposure to patient's trauma decreases the stress threshold (Turgoose et al. 2017). In synopsis, the level of stress increases when patients are treated in a humanized way by healthcare professionals. Therefore, they use dehumanization as an antistress strategy (Haque and Waytz 2012).

As stress provokes dehumanization in medicine, it is of interest to consider how that would work, what the role of cognition would be in the process, and how experiencing stress would affect the evaluation of dehumanizing behavior. In the present study, I addressed these issues in two groups of medical nurses: the experimental group that sat for professional exam just after being surveyed, where examination stress was the experimental intervention, and the control group whose emotional state was neutral. The study hypotheses were the following: (i) stressful situation would induce negative changes in the current affective state, (ii) experiencing stress would facilitate the acceptance of dehumanizing behaviors, and (iii) attitude conducive to dehumanization behavior would correlate with a lower level of empathy and a higher level of neuroticism and psychoticism.

2 Methods

2.1 Participants and Protocol

This research was an empirical interventional study estimating the influence of an experimental intervention without a randomized assignment. The experimental group consisted of nurses taking part in a professional class ending up with examination before being accepted by the Medical Commission. Participants were informed that the exam would be delayed for 45 min. They were then asked to fill in the survey questionnaires. Thus, survey was conducted without the antecedent notice in the emotion-strained condition immediately before the exam. In the control group, participants were surveyed in an emotion–neutral setting, with no exam in the offing.

The study involved 96 nurses working in various wards of hospitals in the cities of Opole and Cracow in Poland. There were 49 nurses in the experimental group and 47 age-matched nurses in the control group. The mean nurses' age was 38 ± 11 years and 40 ± 12 years, respectively. There were no inter-group differences in job seniority either: 15 ± 11 and 18 ± 13 years, respectively (Table 1).

2.2   Psychometric Tools

The evaluation of dehumanizing behavior was conducted with the Scale of Behavioral Indicators of Patient's Dehumanization (SBIPD) of Głębocka and Wilczek–Rużyczka (2016). The questionnaire consists of two subscales: patent and latent dehumanization. The patent dehumanization subscale describes nine short situations which provide obvious and clear evidence of either subjective or objective attitudes of medical staff toward patients. The latent dehumanization subscale includes ten items containing descriptions of situations with hidden signs of dehumanization. All the situations are rated on a five-point scale from 1, very bad, to 5, very good.

The nurses were asked to decide whether a given situation was real or fictitious, although all of the situations actually occurred in real life. The indicator of questionnaire's reliability was satisfactory (Cronbach's alpha = 0.77).

Mood was evaluated with the UMACL Mood Adjective Checklist of Matthews et al. (1990). This scale measures self-reported mood and current emotional state divided into three dimensions: (i) tense arousal, (ii) energetic arousal, and (iii) hedonic tone. Empathy was evaluated with the interpersonal reactivity index (IRI) of Davis (1980). The questionnaire contains four subscales: (i) perspective taking (PT) evaluating a spontaneous adoption of the psychological point of view of others, (ii) fantasy (F) evaluating the tendencies to transpose themselves imaginatively into the feelings and actions of fictitious characters, (iii) empathic concern (EC) evaluating the strength of attention and care of others, and (iv) personal distress (PD) describing the self-oriented feelings of personal anxiety reflecting the suffering of others.

Finally, personality traits were explored with the personality questionnaire (EPQ-R) of Eysenck et al. (1985). This scale consists of four subscales: (i) neuroticism, characterized by

Table 1 Inter-group comparisons for study dimensions

Variable	Experimental group	Control group
Age (years)	38 ± 11	40 ± 12
Job seniority (years)	15 ± 11	18 ± 13
Tense arousal (UMACL)	2.6 ± 0.4	2.2 ± 0.4*
Hedonic tone (UMACL)	2.8 ± 0.3	2.7 ± 0.3
Energetic arousal (UMACL)	2.8 ± 0.5	3.0 ± 0.4
Perspective taking (IRI)	2.4 ± 0.6	2.3 ± 0.5
Fantasy (IRI)	1.7 ± 0.7	1.8 ± 0.7
Empathic concern (IRI)	2.2 ± 0.5	2.3 ± 0.5
Personal distress (IRI)	4.3 ± 0.4	4.3 ± 0.5
Lying scale (EPQ-R)	12.2 ± 3.1	12.2 ± 3.7
Neuroticism (EPQ-R)	14.9 ± 5.0	15.7 ± 4.7
Extraversion (EPQ-R)	14.2 ± 3.5	14.5 ± 4.2
Psychoticism (EPQ-R)	10.0 ± 3.4	10.4 ± 3.5
Patent dehumanization (SBIPD)	1.6 ± 0.5	1.4 ± 0.3*
Latent dehumanization (SBIPD)	1.6 ± 0.6	1.5 ± 0.4

Data are means ±SD. (UMACL) The Mood Adjective Checklist, (IRI) The Interpersonal Reactivity Index Assessing Empathy, (EPQ-R) The Eysenck Personality Questionnaire-Revised, (SBIPD) Scale of Behavioral Indicators of Patient's Dehumanization
*means $p < 0.05$

instability, nervousness, and anxiety; (ii) extraversion, signalized sociability, impulsiveness, hedonism, and a tendency for aggressive behavior, (iii) psychoticism, considered an antisocial trait of personality, e.g., hostility and lack of empathy; and (iv) lying scale.

Data were presented as means $\pm$SD. For normally distributed data, statistical differences between the two groups were evaluated with multivariate analysis of variance (MANOVA) for repeated measures, followed by post hoc Scheffe's test. Non-normal distributed data were evaluated with the Kruskal–Wallis test by ranks.

3 Results

The UMACL scores of current mood and emotional state subjected to multivariate analysis of variance (MANOVA) for repeated measures and against the group factor (experimental vs. control). There were significant inter-group differences for main effects (F(1,93) = 4.79; p = 0.03 η^2 = 0.05) and for interaction effects (F(2,19) = 11.10; p = 0.00003, η^2 = 0.10). Scheffe's post hoc analysis showed significant inter-group differences for tense arousal, where the experimental group scored higher than the control one: 2.56 $\pm$ 0.36 vs. 2.24 $\pm$ 0.38 points, respectively; p < 0.001. That result confirmed the effectiveness of the experimental manipulation used in that the expectation of examination adversely affected the emotional state of the nurses as they were more stressed (afraid or anxious) than the nurses in the control group (Table 1).

An analysis of dehumanizing behaviors was split into patent dehumanization and latent dehumanization of SBIPD questionnaire. Both components failed to meet the normal distribution assumptions and were evaluated with the Kruskal–Wallis test by ranks. There was a significant difference between the experimental and control groups in case of patent dehumanization (mean scores of 1.6 $\pm$ 0.5 and 1.4 $\pm$ 0.3 points, respectively; p < 0.05) but not in latent dehumanization (mean scores of 1.6 $\pm$ 0.6 and 1.5 $\pm$ 0.4 points, respectively; p > 0.05) (Table 1). Despite that all of the situations included in the patent and latent dehumanization scales were real, participants from both groups believed that only 60% of the situations were true. Both groups expressed a negative attitude toward dehumanization behaviors depicted in the real-life accounts, so that the questionnaire scores were on the low side. Nonetheless, participants in the experimental group appeared significantly more accepting and tolerant for the patent dehumanization behaviors than those in the control group.

The IRI scores evaluating aspects of empathy and personality traits were subject to multivariate analysis of variance (MANOVA) for repeated measures. There were no significant differences between the experimental and control groups in the global empathy score (F(3,28) = 0.74, p = 0.52, η^2 = 0.003), with the mean scores of 2.1 $\pm$ 0.5 and 2.1 $\pm$ 0.5 points, respectively; p > 0.05. Nor were there any appreciable differences between the two groups among the dimensions of perspective taking (PT), fantasy (F), empathic concern (EC), and personal distress (PD). Likewise, there were no significant inter-group differences for personality traits evaluated with the EPQ-R questionnaire such as neuroticism, extraversion, and psychoticism (F (4,96) = 0.31, p = 0.860). Of note, the mean scores of 12.0 points in the lying subscale of EPQ-R in both experimental and control groups speak for a good reliability of the outcomes obtained (Table 1).

Finally, correlation analysis was conducted for each group to explore the associations among dehumanizing behaviors, current mood, empathy, and personality features. There appeared some moderate associations, different in either group. In the experimental group, tense arousal (UMACL) associated with fantasy (IRI) (r = 0.32, p < 0.05) and empathy (IRI) (r = 0.32, p < 0.05), whereas in the control group, tense arousal (UMACL) associated with personal distress (IRI) (r = 0.32, p < 0.05). Patent and latent dehumanization behaviors (SBIPD) failed to associate with variables appraising mood and empathy (UMACL and IRI). However, there was an association between patent dehumanization and neuroticism (r = 0.36, p < 0.05) in the experimental group.

4   Discussion

In this study, the experimental manipulation consisted of the use of a stress provoking condition, such as anxiety linked to the ongoing essential examination for getting licensed for nursing profession, to evaluate the possible effects of enhanced level of stress on dehumanizing behaviors in medical profession. To start with, I verified the effectiveness of the manipulation in inducing a negative change in the current affective state, as the expectation of examination increased the level of tense arousal, which is characteristic of stress. The major finding was that anxiety-prone condition enhanced the prospective nurses' attitude toward patent dehumanizing behaviors. They became more prone to estimate such behaviors as acceptable, which unmasked a subconscious negative trend. Patent dehumanization is defined as obvious and clear evidence of humiliating and harmful attitudes of medical staff toward patients. Verbally expressed examples of such attitudes may be as follows:

(i) *A patient visits a surgeon to remove early necrosis of a lower limb. The doctor examines the patient, causing pain. He then tells the patient that the real pain would be felt during surgery and that the patient would not receive anesthetic.*

(ii) *An oncology nurse enters the corridor and tells the patients "Metastatic cancer to the right, non-metastatic cancer to the left"* (Głębocka and Wilczek–Rużyczka 2016).

Although this study provides evidence only for the impact of stress on cognitive aspects of attitude, there is a consistent impression that the acceptance of dehumanization may also affect behavior of medical staff. The finding of hard evidence of dehumanizating behaviors in medicine is a difficult task, since tolerance for dehumanization, as much as inhuman attributions of patients, has mostly to do with subconscious mechanisms (Capozza et al. 2016).

In some previous studies, medical staff has perceived their own group, be it nurses or physicians, as having more uniquely human traits, while less uniquely human traits have been ascribed to patients. It follows that patients were assigned a lower human status. Ascribing patients in this way is tantamount to functional dehumanization, possibly an expression of defensive strategy of coping with stress among medical staff (Trifiletti et al. 2014; Haque and Waytz 2012). There is also evidence that patients' dehumanization associates with lower burnout and a higher level of self-evaluated efficacy and competence (Vaes and Muratore 2013). Other studies have examined the role of the individual's mental capacity to recognize other people's mental states for the relationship between dehumanization and lower self-reported stress. The results show that dehumanization associates with lower burnout in medical staff who has a lower capacity to read other people's minds. This mechanism of coping with stress, consisting of assigning patients to a lower human status, is used by the staff who is unable to make out other people's emotions and thoughts (Capozza et al. 2016; Decety 2011; Han et al. 2008). In fact, dehumanization in medicine seems a strategy to avoid the affective costs of helping (Bakker et al. 2001; Pearlman and Saakvitne 1995).

The evidence above discussed points to the presence of moderators between dehumanization and stress, such as regard patients as non-uniquely human or the incapacity to recognize other people's mental states (Fig. 1). A more positive evaluation of dehumanizing behavior under stress, noticed in the present study, relates possibly more to the latter. However, a number of further research questions arise, which require clarification, such as: (i) why does stress influence the appraisal of patent dehumanization only; (ii) to what extent stress experienced by medical staff attributes patients with non-uniquely human features, as suggested in the literature (Demoulin et al. 2004; Leyens et al. 2001); and (iii) does stress determine the individual's capacity to recognize patients' mental states.

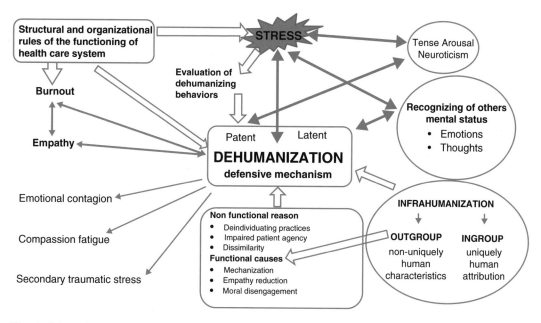

Fig. 1 Schematic representation of psychological moderators influencing the relationship between stress and dehumanizing behaviors

One of the working presumptions in this study was that a more easily exculpated attitude toward dehumanization behaviors could relate with a lower level of empathy and higher levels of neuroticism and psychoticism. However, no significant relationship between relative tolerance to dehumanizing behaviors and personality features was found in the control group, and there was only a moderate relationship between patent dehumanization and neuroticism in the experimental group. That could stem from the stress-enhanced study condition, confirming previous observations that anxiety and fear correspond with enhanced tolerance for dehumanization (Turgoose et al. 2017). Thus, dehumanization in medicine does not necessarily result from caregivers' personality traits or bad intentions. It rather seems a by-product of specific interpersonal relationships and hospital demands strongly connected with the structural and organizational rules of healthcare functioning and with the functional psychological demands intrinsic to medical profession (Haque and Waytz 2012).

Overall, dehumanization in medicine is understood as a dysfunctional strategy of coping with stress. On the one hand, dehumanized patients are more vulnerable to unsatisfactory relationships with medical staff, e.g., stigmatization or prejudice, which can adversely affect compliance with medical recommendations. On the other hand, dehumanized patients can be perceived by physicians as less capable of following medical recommendations. Using dehumanization to reduce stress may have beneficial effects on caregivers' well-being, but not on their relationships with patients, or on patients' physical and psychological well-being. Dehumanization connected with emotional exhaustion results from stress experienced at work, which may also be influenced by poor working conditions, work overload, long working hours, inadequate gratification, or conflicting interpersonal relationships. The best way to fight dehumanization among medical staff is to reduce the level of stress by eliminating pathological work-related issues (Bakker et al. 2005; Shirom 1989).

Acknowledgments The study was co-financed from the funds allocated to the statutory activity of the Faculty of Psychology and Humanities of the Andrzej Frycz Modrzewski University, Cracow, Poland; project WPiNH/DS/5/2018.

Conflicts of Interest The author declares no conflicts of interest in relation to this article.

Ethical Approval All procedures performed in studies involving human participants were in accordance with the ethical standards of the institutional and/or national research committee and with the 1964 Helsinki declaration and its later amendments or comparable ethical standards. The study protocol was approved by the Research Project Committee of the Faculty of Psychology and Humanities, The Andrzej Frycz Modrzewski University in Cracow, Poland.

Informed Consent Subjects participated in the study on a voluntary and anonymous basis. Informed consent was obtained from all individual participants included in the study.

References

Bakker AB, Schaufeli WB, Sixima HJ, Bosveld W (2001) Burnout contagion among general practitioners. J Soc Clin Psychol 20:82–98

Bakker AB, Le Blanc PM, Schaufeli WB (2005) Burnout contagion among intensive care nurses. J Adv Nurs 51:276–287

Capozza D, Trifiletti E, Vezzali L, Favara I (2013) Can intergroup contact improve humanity attributions? Int J Psychol 48:527–541

Capozza D, Falvo R, Boin J, Colledani D (2016) Dehumanization in medical contexts: an expanding research field. Test Psychom Methodol Appl Psychol 23 (4):545–559

Costello K, Hodson G (2010) Exploring the roots of dehumanization: the role of animal–human similarity in promoting immigrant humanization. Group Process Intergroup Relat 13:3–22

Davis MH (1980) A multidimensional approach to individual differences in empathy. JSAS Cat Sel Doc Psychol 10:85

de Zulueta P (2013) Compassion in 21st century medicine: is it sustainable? Clin Ethics 8(4):119–128

Decety J (2011) Dissecting the neural mechanisms mediating empathy. Emot Rev 3:92–108

Demoulin S, Leyens JP, Paladino MP, Rodríguez–Torres R, Rodríguez–Peréz A, Dovidio JF (2004) Dimensions of "uniquely" and "non–uniquely" human emotions. Cognit Emot 18:71–96

Eysenck SBG, Eysenck HJ, Barrett P (1985) A revised version of the psychoticism scale (PDF). Personal Individ Difffer 6(1):21–29

Figley CR (2002) Compassion fatigue: psychotherapists' chronic lack of self–care. J Clin Psychol 58:1433–1441

Głębocka A (2017) The relationship between burnout syndrome among the medical staff and work conditions in the polish healthcare system. Adv Exp Med Biol 179:1–10

Głębocka A, Wilczek–Rużyczka E (2016) Dehumanizing behaviors toward patients: medical staff perspective. Czasopismo Psychologiczne 22(2):1–8. (Article in Polish)

Han S, Fan Y, Mao L (2008) Gender difference in empathy for pain: an electrophysiological investigation. Brain Res 1196:85–93

Haque OS, Waytz A (2012) Dehumanization in medicine: causes, solutions, and functions. Perspect Psychol Sci 7(2):176–186

Haslam N, Stratemeyer M (2016) Recent research on dehumanization. Curr Opin Psychol 11:25–29

Leyens JP, Rodríguez–Pérez A, Rodríguez–Torres R, Gaunt R, Paladino MP, Vaes J, Demoulin S (2001) Psychological essentialism and the differential attribution of uniquely human emotions to ingroups and outgroups. Eur J Soc Psychol 31:395–411

Leyens JP, Demoulin S, Vaes J, Gaunt R, Paladino MP (2007) Infra–humanization: the wall of group differences. Soc Issues Policy Rev 1:139–172

Matthews G, Jones DM, Chamberlain AG (1990) Refining the measurement of mood: the UWIST mood adjective checklist. Br J Psychol 81:17–42

Pearlman LA, Saakvitne KW (1995) Treating therapists with vicarious traumatization and secondary traumatic stress disorders. In: Figley C (ed) Compassion fatigue: coping with secondary–traumatic stress disorder in those who treat the traumatized. Brunner Mazel, New York

Shirom A (1989) Burnout in work organizations. In: Cooper CL, Robertson I (eds) International review of industrial and organizational psychology. Wiley, Chichester

Trifiletti E, Di Bernardo GA, Falvo R, Capozza D (2014) Patients are not fully human: s nurse's coping response to stress. J Appl Soc Psychol 44:768–777

Turgoose D, Glover N, Barker C, Maddox L (2017) Empathy, compassion fatigue and burnout in police officers working with rape victims. Traumatology 23 (2):205–213

Vaes J, Muratore M (2013) Defensive dehumanization in the medical practice: a cross–sectional study from a health care worker's perspective. Br J Soc Psychol 52:180–190

Vaes J, Leyens JP, Paladino MP, Miranda MP (2012) We are human, they are not: driving forces behind outgroup dehumanisation and the humanisation of the ingroup. Eur Rev Soc Psychol 23:64–106

Zimbardo PG (1970) The human choice: individuation, reason and order versus deindividuation, impulse, and chaos. In: Arnold WJ, Levine D (eds) Nebraska symposium on motivation. University of Nebraska Press, Lincoln

Adv Exp Med Biol - Clinical and Experimental Biomedicine (2019) 4: 105–114
https://doi.org/10.1007/5584_2018_296
© Springer Nature Switzerland AG 2018
Published online: 30 November 2018

Biological and Social Determinants of Maximum Oxygen Uptake in Adult Men

Stanisław B. Nowak, Andrzej Jopkiewicz, and Paweł Tomaszewski

Abstract

The maximum rate of O_2 uptake (VO_2max) is one of the most important positive indicators of health. While the VO_2max decreases with age, reducing the capacity for physical effort, it can be considerably upregulated through optimal environmental interventions, including systematic physical activity. This study seeks to determine variations in the cardiorespiratory function, estimated from the level of VO_2max, in 798 employed men aged 20–59, according to biological (age, physical activity, body mass index (BMI), and limb muscle strength and agility) and social (place of residence, education, occupation, economic status, and smoking) predictors. We found that the variables abovementioned, with the exception of smoking and hand strength, were significant predictors of VO_2max in univariate logistic regression, with age (OR = 0.52; 95%CI 0.47–0.57) and BMI (OR = 0.91; 95%CI 0.90–0.93) having the greatest effect on VO_2max. The additional predictors, established in multivariate analysis, were the place of residence, education, and hand and arm strength. The multivariate model was fairly well-fitted (Nagelkerke $r^2 = 0.54$) and had a satisfactory prognostic value, with over 80% of cases classified correctly. Social variance in the VO_2max makes it desirable to develop and implement the intervention programs with physical activity dedicated for men, especially men who are over the age of 50 years and have an excessive body mass, as this could reduce the risk of disorders and help improve the quality of life and workplace effectiveness of this group.

Keywords

Cardiorespiratory function · Health indicators · Maximum oxygen uptake · Physical activity · Quality of life

S. B. Nowak (✉)
Department of Physical Education, Kazimierz Pulaski University of Technology and Humanities in Radom, Radom, Poland
e-mail: snowak@uthrad.pl

A. Jopkiewicz
Department of Auxology, Jan Kochanowski University in Kielce, Kielce, Poland

P. Tomaszewski
Department of Biometry, Jozef Pilsudski University of Physical Education in Warsaw, Warsaw, Poland

1 Introduction

In today's era, the biology of the human body is increasingly at risk due to a lack of physical activity and the consequences thereof, in particular, resulting in overweight and obese condition. We are observing a steady shift from an active lifestyle in the external environment, which has been natural for the human race, toward a sedentary lifestyle in the enclosed spaces (O'Keefe et al. 2010; Malina and Litle 2008), even though the beneficial effect of aerobic exercise on the body is an evolutionary acquisition (Rowe et al. 2014). Endurance-based exercise, involving prolonged walking and running, has played a key role in human evolutionary history, and our species is distinguished from other primates in this respect (Mattson 2012).

Despite the fact that aerobic activity involves a high energy expenditure and fatigue, nature has programed the human body to feel pleasure during and after exercise, which is often expressed as a feeling of "runner's high". This is related to the activity of the reward brain center, resulting from an increase in the amount of endocannabinoids that lead to mood improvement. Thus, natural selection, through the endocannabinoid system, has helped motivate human beings to perform physical exercise, which in turn has not only ensured human survival but has also been crucial in anthropogenesis (Raichlen et al. 2012).

Today, the role of physical exercise as a predictor of survival has considerably decreased, and even the neurobiological rewards are not enough to encourage physical activity. Nonetheless, regular aerobic exercise is indisputably related to good health and longevity; whereas the opposite is true for a sedentary lifestyle, which leads to various health issues and premature mortality (Després 2016; Laukkanen et al. 2016; Arem et al. 2015; Gebel et al. 2015; Hupin et al. 2015; Blair 2009). A shift from a traditionally active lifestyle to a sedentary one is a global phenomenon. Types of physical activity requiring large energy expenditures (heavy physical labor and traveling on foot) have been replaced with low-energy forms of activity (office work and mechanized travel) (Katzmarzyk and Mason 2009).

The World Health Organization classifies physical inactivity as the fourth leading cause of global mortality and the primary cause of many chronic disorders (WHO 2009). The term physical activity transition has been coined to underline

this harmful tendency, which is particularly dangerous to the health of children and youth (Katzmarzyk and Mason 2009). The yearly health costs resulting from a low level of physical activity are estimated to exceed $67 billion globally, and a sedentary lifestyle causes about 5 million deaths *per* year (i.e., almost 10% of the deaths not resulting from violence) (Ding et al. 2016). Hallal et al.'s (2012) analysis of the data collected from 122 countries shows that over 31% of adolescents and adults aged over 15 years are physically inactive. The inactivity is more common in wealthy countries and among women and elderly persons, and it factors in the development of noncommunicable diseases (Dumith et al. 2011). Age, sex, health, obesity, self-efficacy, and motivation are other factors associated with the level of physical activity. The evidence shows that the availability of a sport and recreation infrastructure close to one's place of residence has a positive causal relationship with the level of physical activity among both youths and adults (Smith et al. 2017; Bauman et al. 2012).

One of the best measures of functional efficiency is the maximal oxygen uptake (VO_2max). Oxygen requirement of the working muscles is an objective indicator of the cardiorespiratory function (CRF) related to habitual physical activity. The VO_2max is an independent, diagnostic, and prognostic health indicator (Lee et al. 2010) that determines predispositions for prolonged aerobic exercise (Kenney et al. 2015; Araújo et al. 2013; Hawkins et al. 2007). Oxygen plays a key role in CRF as it is needed for the conversion of adenosine triphosphate into energy in the muscle cells. Consequently, the greater the oxygen uptake, the more energy can be produced.

VO_2max represents the highest rate at which oxygen can be transported and used during aerobic exercise. It denotes the maximum volume of oxygen that a person can process *per* minute. Except the highly trained athletes, the contemporary global population displays a lower VO_2max than it could have had (Powell et al. 2011; Sagiv et al. 2007). After the age of about 30 years, the VO_2max begins to decrease consistently by about 0.5–0.6 ml/kg/min *per* year, due primarily to evolutionary changes that take place in the cardiopulmonary system and in the muscles. A value approaching 50 ml/kg/min is considered satisfactory for middle-aged men. The lowest relative value of the VO_2max, required for a full locomotive independence, is about 15 ml/kg/min.

Social determinants of health play a central role in morbidity and mortality among men and contribute to a heath-wise gender disparity (Leone and Rovito 2013; Lee et al. 2010, 2011). Men tend to undertake riskier health behaviors and are more likely to avoid prophylactic care than women do. These differences concern not only the men themselves but also their relatives and may have a negative effect on their participation in the job market (Giorgianni et al. 2013; Kwan et al. 2012; Evans et al. 2011). Therefore, the main aim of this study was to assess differences in CRF based on the VO_2max of working men aged 20–59 years, according to social and biological predictors.

2   Methods

The study was conducted at several workplaces in the Swiętokrzyskie province in Poland during the spring of 2015. Participants consisted of a cohort of 798 men, stratified into 4 age groups: 20–29, 30–39, 40–49, and 50–59 years. The basic inclusion criterion was a combination of non-probability and random sampling, with priority given to randomly selected workplaces that employed mostly men and to divisions of the provincial vocational training center that conducted training for persons in various occupations who lived in both urban and rural areas and who had different levels of education. The inclusion criteria were a lack of health contraindications for performing a voluntary physical exercise workout. During a qualification interview, the participants were instructed about the scope of the study and were informed that they could opt out at any stage without providing a reason. All measurements were taken before noon, and the workouts were preceded by a warm-up.

The independent variables characterizing the social variation among the study participants were determined using a categorized interview, and they comprised of place of residence (large city, small city, or village), level of education (higher, secondary, or vocational), occupation (involving intellectual or physical labor), financial status (low, average, or good), and smoking habits (non-smoker, occasional smokers, or smokers over six cigarettes *per day*) (Table 1). The biological variables were age as a continuous categorized variable (20–29, 30–39, 40–49, or 50–59 years), BMI (kg/m^2) as a continuous and categorized variable (normal body mass, overweight, or obesity) according to the WHO (2008) classification, and the interview-based level of free-time physical activity categorized into low, moderate, or high. Physical activity was assessed on the basis of the number of days *per* week in which the participant performed at least 30 min of intense physical exercise (as a one-time exercise or a sum of the exercise periods for at least 10 min), i.e., the exercise that would create a feeling of tiredness. The applied measure of physical activity was the number of minutes spent on exercise in a week multiplied by the average intensity of exercise expressed in MET (metabolic equivalent). 1 MET corresponds to oxygen uptake during rest and amounts to 3.5 ml O_2/kg/min (Zhang et al. 2003; Araújo et al. 2017). Physical activity was stratified into three groups, according to the method of Lakoski et al. (2011): low (1–449 MET/min/week), moderate (450–749 MET/min/week), and high ($\geq$750 MET/min/week). In addition, several indices of motor ability were taken into account, such as static handgrip strength, static arm strength, dynamic leg strength, and overall agility (a component of speed, involving the coordination of body maneuverability), where:

- Static grip strength of the dominant hand was measured to an accuracy of 1 kg using a hydraulic manual dynamometer.
- Dynamic arm strength was measured based on the number of flexions and extensions of the arms performed within 30 s during exercise with a front support.

Table 1 Characteristics of the study participants according to aerobic capacity (VO$_2$max); numbers (%) of participants and means $\pm$SD are provided for the qualitative and continuous variables, respectively

Variable	VO$_2$max below median (<33.0 ml/kg/min) (n = 400)	VO$_2$max above median (>33.0 ml/kg/min) (n = 398)	Total (n = 798)
Age (years)	44.5 ± 10.1	34.8 ± 9.9	39.7 ± 11.1
20–29	42 (10.5)	128 (32.2)	170 (21.3)
30–39	74 (18.5)	141 (35.4)	215 (26.9)
40–49	136 (34.0)	99 (24.9)	235 (29.5)
50–59	148 (37.0)	30 (7.5)	178 (22.3)
BMI (kg/m^2)[a]	27.1 ± 2.8	23.5 ± 2.2	25.3 ± 3.1
Normal	76 (19.0)	299 (75.5)	375 (47.1)
Overweight	275 (68.7)	96 (24.2)	371 (46.6)
Obese	49 (12.3)	1 (0.3)	50 (6.3)
Place of residence			
Village	76 (19.0)	136 (34.2)	212 (26.6)
Small city	92 (23.0)	82 (20.6)	174 (21.8)
Large city	232 (58.0)	180 (45.2)	412 (51.6)
Education			
Vocational	138 (34.5)	178 (44.7)	316 (39.6)
Secondary	144 (36.0)	125 (31.4)	269 (33.7)
Higher	118 (29.5)	95 (23.9)	213 (26.7)
Occupation			
Physical labor	205 (51.2)	272 (68.3)	477 (59.8)
Intellectual labor	195 (48.8)	126 (31.7)	321 (40.2)
Financial status			
Below average	52 (13.0)	79 (19.8)	131 (16.4)
Average	266 (66.5)	266 (66.6)	531 (66.5)
Good	82 (20.5)	82 (13.6)	136 (17.1)
Smoking			
Non-smokers	163 (40.8)	159 (39.9)	322 (40.4)
Occasional smokers	86 (21.5)	70 (17.6)	156 (19.5)
Smokers >6 cigarettes/day	151 (37.7)	169 (42.5)	320 (40.1)
Physical activity			
Low	272 (68.0)	234 (58.8)	506 (63.4)
Moderate	102 (25.5)	122 (30.7)	224 (28.1)
High	26 (6.5)	42 (10.5)	68 (8.5)
Static hand strength (kg)	43.1 ± 8.9	42.8 ± 9.5	42.9 ± 9.2
Dynamic arm strength (n)	13.5 ± 8.3	17.8 ± 8.4	15.7 ± 8.6
Dynamic leg strength (cm)	162.5 ± 36.0	183.6 ± 32.3	173.0 ± 35.8
Agility (s)	28.5 ± 3.3	27.6 ± 3.1	28.0 ± 3.2
VO$_2$max (ml/kg/min)	27.5 ± 4.2	40.4 ± 6.5	34.0 ± 8.5
VO$_2$max (l/min)	2.2 ± 0.3	2.8 ± 0.4	2.5 ± 0.5

[a]n = 796; two underweight persons were excluded from analysis

- Explosive strength of legs was measured based on a standing long jump (cm).
- Agility was measured based on a zigzag run within a 5 × 3 m area (three laps); the participants had to move around five poles (four in the corners and one in the middle), while the running time was measured to an accuracy of 0.1 s, with the better result from two attempts taken into account.

VO$_2$max was determined indirectly using the Astrand test performed on a Monark LC6 cycle ergometer (Monark Exercise AB, Vansbro, Sweden). This test was divided into two stages: 3-min warm-up under a load of 50 W and 5–6-min exercise under a submaximal workload of 100–150 W, performed until the participant's heart rate, measured with a cardiometer, stabilized at 130–160 beats/min. The VO$_2$max was then calculated using the Astrand and Ryhming (1954) tables, according to the heart rate, workload, body mass, and age. The participants' characteristics and VO$_2$max levels are presented in Table 1.

The results were expressed as means ±SD or numbers and corresponding percentages. Relationships between the qualitative variables were assessed using a chi-squared test; Cramér's V was used as a measure of effect size. Groups with a VO$_2$max below and above the median (Me) of 33.0 ml/kg/min were selected for the purpose of a logistic regression. Univariate and multivariate models of logistic regressions were used to assess the probability of the occurrence of a higher level of VO$_2$max (dependent variable) according to the age, BMI, and social and physical variables (independent variables). Multivariate analyses were conducted using the backward stepwise procedure, with age and BMI analyzed in the categorized forms. For the independent variables, odds ratios (OR) with 95% confidence intervals (CI) were calculated, and the Nagelkerke r^2 was estimated as a measure of effect size. The statistical significance was assumed at $\alpha = 0.05$ for all analyses.

3 Results

Table 2 presents the odds ratios for the occurrence of a higher VO$_2$max, estimated through the univariate logistic regression analysis. Significant predictors of the VO$_2$max were found to comprise nearly all the variables, with the exception of smoking and static hand strength. The factors that had the largest effect on the VO$_2$max were the age and BMI ($p < 0.001$), both with inverse correlations. An increase in BMI by 1 point correlated with a nearly twice as low probability of a higher VO$_2$max (OR = 0.52; 95%CI 0.47–0.57), and an increase in age by 1 year correlated with a 10% lower probability of a higher VO$_2$max (OR = 0.91; 95%CI 0.90–0.93). Among the oldest participants and the obese participants, the probability of a higher VO$_2$max was over 14 times lower (OR = 0.07; 95%CI 0.04–0.11) and 100 times lower, respectively, compared to the younger participants and the participants with a correct body mass. Only 1 in 50 of the obese participants showed a higher VO$_2$max, compared to about 75% of the participants with a correct BMI (Table 2). Furthermore, the variables of living in a city and performing intellectual labor correlated with a nearly twice as low probability of a higher VO$_2$max (OR 0.5 and 0.49, respectively), compared to the participants that lived in rural areas and performed physical labor. In turn, a higher level of physical activity and better results in the arm flexion test (dynamic arm strength) and the standing long jump test (dynamic leg strength) increased the probability of a higher VO$_2$max. For the persons who declared a high level of physical activity, the probability of a higher VO$_2$max was nearly twice as high (OR = 1.88; 95%CI 1.12–3.16) compared to those who declared low levels of physical activity. An improvement in the results of the arm flexion test and the standing long jump test by one unit correlated with a 7% and 2% higher probability of a higher VO$_2$max, respectively.

The multivariate logistic regression analysis conducted with the backward stepwise method

Table 2 Odds ratios of a higher VO_{2max} (> median VO_{2max}) according to the biosocial variables studied, based on the univariate models of logistic regression (n = 798)

Variable	OR	95%CI	p	Nagelkerke r^2
Age	0.91	0.90–0.93	<0.001	0.25
20–29	1			0.24
30–39	0.63	0.40–0.98	0.04	
40–49	0.24	0.15–0.37	<0.001	
50–59	0.07	0.04–0.11	<0.001	
BMI	0.52	0.47–0.57	<0.001	0.48
Normal	1			0.41
Overweight	0.09	0.06–0.12	<0.001	
Obese	0.01	0.00–0.04	<0.001	
Place of residence				
Village	1			0.04
Small city	0.50	0.33–0.75	<0.001	
Large city	0.43	0.31–0.61	<0.001	
Education				
Vocational	1			0.01
Secondary	0.67	0.49–0.93	0.02	
Higher	0.62	0.44–0.89	0.01	
Occupation				
Physical labor	1			0.04
Intellectual labor	0.49	0.37–0.65	<0.001	
Financial status				
Below average	1			0.02
Average	0.66	0.44–0.97	0.03	
Good	0.43	0.27–0.71	<0.001	
Smoking				
Non-smokers	1			<0.01
Occasional smokers	0.83	0.57–1.22	0.36	
Smokers >6 cigarettes/day	1.15	0.84–1.56	0.38	
Physical activity				
Low	1			0.01
Moderate	1.39	1.01–1.91	0.04	
High	1.88	1.12–3.16	0.02	
Static hand strength	1.00	0.98–1.01	0.594	<0.001
Dynamic arm strength	1.07	1.05–1.09	<0.001	0.09
Dynamic leg strength	1.02	1.01–1.02	<0.001	0.12
Agility	0.92	0.88–0.96	<0.001	0.03

confirmed the correlation between the VO_2max and age, BMI, education, and place of residence (Table 3). In addition, the dynamic arm strength was found to be a significant predictor of VO_2max (OR = 1.04; 95% CI 1.02–1.07), and the static hand strength showed a negative correlation with the VO_2max (OR = 0.96; 95% CI 0.94–0.99) in contrast to the results of the univariate analysis. The other variables were insignificant, and they were not included in the final model. They correlated strongly with the variables included in the model and were thus redundant: physical activity correlated with age ($\chi^2_{6,798}$ = 50.9; p < 0.001; Cramér's V = 0.18) and education ($\chi^2_{4,798}$ = 40.3; p < 0.001; Cramér's V = 0.16); occupation correlated with the place of residence ($\chi^2_{2,798}$ = 78.1; p < 0.001; Cramér's V = 0.31); and financial status correlated with the level of

Table 3 Odds ratios of a higher VO_2max, (> median VO_{2max}) according to the biosocial variables studied, based on the multivariate analysis of logistic regression (n = 798)

Variable	OR	OR 95%CI	p	Nagelkerke r^2
Age				0.54
20–29	1			
30–39	0.74	0.41–1.30	0.29	
40–49	0.30	0.17–0.53	<0.001	
50–59	0.08	0.04–0.16	<0.001	
BMI				
Normal	1			
Overweight	0.12	0.08–0.17	<0.001	
Obese	0.01	0.00–0.04	<0.001	
Place of residence				
Village	1			
Small city	0.58	0.32–1.05	0.07	
Large city	0.52	0.32–0.85	0.01	
Education				
Vocational	1			
Secondary	0.47	0.29–0.76	<0.002	
Higher	0.38	0.22–0.65	<0.001	
Static hand strength	0.96	0.94–0.99	<0.004	
Dynamic arm strength	1.04	1.02–1.07	<0.002	

education ($\chi^2_{4,798} = 80.1$; p < 0.001; Cramér's V = 0.22). Overall, the estimated multivariate model was found to be fairly well-fitted (Nagelkerke $r^2 = 0.54$) and to have a satisfactory prognostic value as over 80% of the cases were correctly classified.

4 Discussion

In this study, we found that significant predictors of the aerobic capacity, based on VO_2max using, comprised almost all of the variables studied, with the exception of static hand strength. A univariate analysis shows that the largest effect on the VO_2max was exerted by the BMI, age, and, to a slightly lesser extent, the place of residence and education. These results are in line with the research conducted at the Cooper Clinic in Dallas, Texas, between 2000 and 2010 that investigated the modified and unmodified determinants of the cardiorespiratory function (CRF) (Lakoski et al. 2011). In that research, the strongest clinical factors were determined using a linear regression model. The BMI, age, gender, and physical activity have been found the most important factors related to CRF, accounting for 56% of the variation. Akin to the present study, the BMI was the strongest clinical risk factor related to CRF, alongside unmodified risk factors, such as the participant's age or gender. For the participants with a similar level of physical activity, those with a normal BMI had a higher CRF compared to obese persons. Overall, the data suggest that obesity may negate the benefits of physical activity, even in a healthy population of men and women.

The specifics of a steep decline in peek aerobic capacity in persons undergoing training have been described by Sagiv et al. (2007). Those authors demonstrate that the rate of the muscle strength and aerobic capacity decline, indexed as the peak VO_2, are key from the viewpoint of quality of life and functional independence. The decline is not constant in healthy adults, as may be assumed from cross-sectional studies showing a 5–10% decline *per* decade of age in untrained persons, but rather, it appreciably increases each decade of age, especially in men. Fleg et al. (2005) have suggested that the rate of decline increases from 3–6% at 20–30 years of age to over 20% *per* decade after the age of 70, and it can also be indexed *per* kilogram of body mass or kilogram of lean body mass.

The effect of social determinants on the VO_2max has been shown in a study that compared the population of Tsimané Indians living in Bolivian Amazonia to a highly industrialized Canadian population, a part of the Tsimané Health and Life History Project carried out between 2002 and 2010. The Indians have a considerably higher VO_2max and, notably, a lower rate of decline than the Canadians do. The Indians' VO_2max is consistent with a high physical activity stemming from farming and contract work (Gurven et al. 2013; Pisor et al. 2013). Living in a rural environment, even the alpine one, and leading a farming-based lifestyle may not be sufficient for a better CRF and physical fitness. Physical activity always needs to have an optimal volume and intensity (Beall et al. 1985). Nonetheless, a lower socioeconomic status of rural population is usually accompanied by a higher level of physical activity and aerobic capacity when compared to better off urban population. This has been confirmed by studies such as the one conducted by Muthuri et al. (2014) among African children, in whom a higher level of physical activity translated into a higher aerobic capacity. That study also demonstrates that a lower education and living in a rural environment associates with a higher VO_2max in men than women. However, the effect of various environmental factors, and rather their aggregation as a single factor can never be solely responsible, should be taken into account when considering different social predictors.

Physical activity improves VO_2max and consequently health. However, different forms of physical activity promote different physiological changes and different levels of health-related benefits (Pimentel et al. 2003). The type, level, volume, and frequency of physical activity are important considerations. According to the recommendations of the US Department of Health and Human Services (2008), adults should perform 500–1000 MET min/week of moderate-to-intense activity. This volume of activity, which corresponds to 150–300 min of fast walk or 75–150 min of jogging, provides major health

benefits. The present study confirm the benefit of physical activity on VO_2max as physically active persons had nearly twice as high a probability of achieving a higher level of VO_2max compared to persons with low physical activity. However, in multivariate analysis, physical activity appeared an insignificant predictor of VO_2max, due likely to its correlation with age and education. Nonetheless, it should be noted that even a low level of physical activity is better than no activity at all, and it may result in health benefits if it is appropriately distributed over time (Powell et al. 2011). Hagströmer et al. (2015) have emphasized that all forms of physical activity, including everyday activities, influence health. They have also demonstrated that the risk of mortality among persons who spend 10 h a day in a sedative lifestyle is over 2.5 times greater than among those who limit their sedative lifestyle to 6.5 h a day. Everyday physical activity for more than half an hour may decrease the risk of death by as much as 50%. Post-training changes in VO_2max are nonlinear and depend on the exercise intensity and duration and on the frequency and length of a training program.

Huang et al. (2016) have determined the duration and parameters of the optimal aerobic training for healthy older persons who lead a sedentary lifestyle. Such persons should participate in a 30–40-week health improvement training program, carried out in 3–4 training sessions a week. Each session should last 40–50 min and have an intensity amounting to 66–73% of the heart rate reserve. The CRF decreases linearly, and its decline increases after the age of 45 years in both men and women. Maintaining a correct body mass, level of physical activity, and not smoking all distinctly contribute to a higher CRF (Jackson et al. 2009).

A decrease in cardiorespiratory function is due primarily to a sedentary lifestyle, which in turn contributes to increased BMI. Undertaking a physical activity is therefore important for health and quality of life in every stage of ontogenesis, and it appears to be indispensable in older age. The present study, conducted in a large cohort of working men, confirms these issues. A limitation of this study is the use of an indirect method of assessing VO$_2$max based on the subject's sub-maximal heart rate. That caused an arbitrary enforcement of the age-specific decline in CRF, which could introduce inaccuracies. In addition, a cross-sectional study design revealed just the cohort effects, whereas, as suggested by Nussey et al. (2008), changes in VO$_2$max could be better explained in longitudinal research due to the issues related to inter-individual heterogenicity and individual aspects of aging.

In conclusion, age and body mass index have the largest effect on cardiorespiratory function, estimated from the level of VO$_2$max, in working men aged 20–59, which was confirmed in multi-variate analysis using the backward stepwise method. We submit that it would be socially desirable to implement an intervention program involving recreational physical activity dedicated to middle-aged men with overweight or obesity, as that could reduce the risk of illness and improve quality of life and occupational effectiveness.

Conflicts of Interest The authors declare no conflicts of interest in relation to this article.

Ethical Approval All procedures performed in the study were in accordance with the ethical standards of the insti-tutional national and/or research committee and with the 1964 Helsinki declaration and its later amendments or comparable ethical standards. The study was approved by the Bioethics Committee of the Faculty of Medicine and Health Sciences of the Jan Kochanowski University in Kielce, Poland.

Informed Consent Informed consent was obtained from all individual participants included in the study.

References

Araújo CG, Herdy AH, Stein R (2013) Maximum oxygen consumption measurement: valuable biological marker in health and in sickness. Arq Bras Cardiol 100(4): e51–e53

Araújo CGS, Castro CLB, Franca JF, Silva CGSE (2017) Aerobic exercise and the heart: discussing doses. Arq Bras Cardiol 108(3):271–275

Arem H, Moore SC, Patel A, Hartge P, Berrington de Gonzalez A, Visvanathan K, Campbell PT, Freedman M, Weiderpass E, Adami HO, Linet MS,

Lee IM, Matthews CE (2015) Leisure time physical activity and mortality: a detailed pooled analysis of the dose-response relationship. JAMA Intern Med 175 (6):959–967

Astrand PO, Ryhming I (1954) A nomogram for calcula-tion of aerobic capacity (physical fitness) from pulse rate during sub–maximal work. J Appl Physiol 7:218–221

Bauman AE, Reis RS, Sallis JF, Wells JC, Loos RJ, Martin BW (2012) Correlates of physical activity: why are some people physically active and others not? Lancet 380(9838):258–271

Beall CM, Goldstein MC, Feldman ES (1985) The physi-cal fitness of elderly Nepalese farmers residing in rug-ged mountain and flat terrain. J Gerontol 40 (5):529–535

Blair SN (2009) Physical inactivity: the biggest public health problem of the 21st century. Br J Sports Med 43(1):1–2

Després JP (2016) Physical activity, sedentary behaviors, and cardiovascular health: when will cardiorespiratory fitness become a vital sign? Can J Cardiol 32 (4):505–513

Ding D, Lawson KD, Kolbe–Alexander TL, Finkelstein EA, Katzmarzyk PT, van Mechelen W, Pratt M, Lancet Physical Activity Series 2 Executive Committee (2016) The economic burden of physical inactivity: a global analysis of major non–communicable diseases. Lancet 388(10051):1311–1324

Dumith SC, Hallal PC, Reis RS, Kohl HW 3rd (2011) Worldwide prevalence of physical inactivity and its association with human development index in 76 countries. Prev Med 53(1–2):4–28

Evans J, Frank B, Oliffe JL, Gregory D (2011) Health, illness, men and masculinities (HIMM): a theoretical framework for understanding men and their health. J Mens Health 8(1):15

Fleg JL, Morrell CH, Bos AG, Brant LJ, Talbot LA, Wright JG, Lakatta EG (2005) Accelerated longitudi-nal decline of aerobic capacity in healthy older adults. Circulation 112(5):674–682

Gebel K, Ding D, Chey T, Stamatakis E, Brown WJ, Bauman AE (2015) Effect of moderate to vigorous physical activity on all–cause mortality in middle–aged and older Australians. JAMA Intern Med 175 (6):970–977

Giorgianni SJ Jr, Porche ST, Williams ST, Matope JH, Leonard BL (2013) Developing the discipline and practice of comprehensive men's health. Am J Mens Health 7(4):342–349

Gurven M, Jaeggi AV, Kaplan H Cummings D (2013) Physical activity and modernization among Bolivian Amerindians. PLoS One 8:1–13

Hagströmer M, Kwak L, Oja P, Sjöström M (2015) A 6 year longitudinal study of accelerometer–measured physical activity and sedentary time in Swedish adults. J Sci Med Sport 18(5):553–557

Hallal PC, Andersen LB, Bull FC, Guthold R, Haskell W, Ekelund U (2012) Global physical activity levels:

surveillance progress, pitfalls, and prospects. Lancet 380(9838):247–257

Hawkins MN, Raven PB, Snell PG, Stray–Gundersen J, Levine BD (2007) Maximal oxygen uptake as a parametric measure of cardiorespiratory capacity. Med Sci Sports Exerc 39(1):103–107

Huang G, Wang R, Chen P, Huang SC, Donnelly JE, Mehlferber JP (2016) Dose–response relationship of cardiorespiratory fitness adaptation to controlled endurance training in sedentary older adults. Eur J Prev Cardiol 23(5):518–529

Hupin D, Roche F, Gremeaux V, Chatard JC, Oriol M, Gaspoz JM, Barthélémy JC, Edouard P (2015) Even a low–dose of moderate–to–vigorous physical activity reduces mortality by 22% in adults aged ≥60 years: a systematic review and meta–analysis. Br J Sports Med 49(19):1262–1267

Jackson AS, Sui X, Hébert JR, Church TS, Blair SN (2009) Role of lifestyle and aging on the longitudinal change in cardiorespiratory fitness. Arch Intern Med 169(19):1781–1787

Katzmarzyk PT, Mason C (2009) The physical activity transition. J Phys Act Health 6:269–280

Kenney WL, Wilmore JH, Costill D (2015) Physiology of Sport and Exercise. 6th Edition with Web Study Guide. Publisher: Human Kinetics Publishers; ISBN-13: 9781450477673

Kwan MY, Cairney J, Faulkner GE, Pullenayegum EE (2012) Physical activity and other health–risk behaviors during the transition into early adulthood: a longitudinal cohort study. Am J Prev Med 42(1):14–20

Lakoski SG, Barlow CE, Farrell SW, Berry JD, Morrow JR Jr, Haskell WL (2011) Impact of body mass index, physical activity, and other clinical factors on cardiorespiratory fitness (from the Cooper Center longitudinal study). Am J Cardiol 108(1):34–39

Laukkanen JA, Zaccardi F, Khan H, Kurl S, Jae SY, Rauramaa R (2016) Long–term change in cardiorespiratory fitness and all–cause mortality: a population–based follow–up study. Mayo Clin Proc 91 (9):1183–1188

Lee D, Artero EG, Sui X, Blair SN (2010) Mortality trends in the general population: the importance of cardiorespiratory fitness. J Psychopharmacol 24(4):27–35

Lee D, Sui X, Artero EG, Lee IM, Church TS, McAuley PA, Stanford FC, Kohl HW 3rd, Blair SN (2011) Long–term effects of changes in cardiorespiratory fitness and body mass index on all–cause and cardiovascular disease mortality in men: the Aerobics Center Longitudinal Study. Circulation 124(23):2483–2490

Leone JE, Rovito MJ (2013) Normative content and health inequity enculturation: a logic model of men's health advocacy. Am J Mens Health 7:243–254

Malina RM, Litle BB (2008) Physical activity: the present in the context of the past. Am J Hum Biol 20 (4):373–391

Mattson MP (2012) Evolutionary aspects of human exercise–born to run purposefully. Ageing Res Rev 11(3):347–352

Muthuri SK, Wachira LM, LeBlanc AG, Francis CE, Sampson M, Onywera VO, Tremblay MS (2014) Temporal trends and correlates of physical activity, sedentary behaviour, and physical fitness among school–aged children in sub–Saharan Africa: a systematic review. Int J Environ Res Public Health 11:3327–3359

Nussey D, Coulson T, Festa–Bianchet M, Gaillard JM (2008) Measuring senescence in wild animal populations: towards a longitudinal approach. Funct Ecol 22:393–406

O'Keefe JH, Vogel R, Lavie CJ, Cordain I (2010) Organic fitness: physical activity consistent with our hunter–gatherer heritage. Phys Sportsmed 38(4):11–18

Pimentel AE, Gentile CL, Tananka H, Seals DR, Gates PE (2003) Greater rate of decline in maximal aerobic capacity with age in endurance–trained than in sedentary men. J Appl Physiol 94(6):2406–2413

Pisor AC, Gurven M, Blackwell AD, Kaplan H, Yetish G (2013) Patterns of senescence in human cardiovascular fitness: VO_2 max in subsistence and industrialized populations. Am J Hum Biol 25(6):756–769

Powell KE, Paluch AE, Blair SN (2011) Physical activity for health: what kind? How much? How intense? On top of what? Annu Rev Public Health 32:349–365

Raichlen DA, Foster AD, Gerdeman GL, Seillier A, Giuffrida A (2012) Wired to run: exercise–induced endocannabinoid signaling in humans and cursorial mammals with implications for the 'runner's high'. J Exp Biol 215:1331–1336

Rowe GC, Safdar A, Arany Z (2014) Running forward: new frontiers in endurance exercise biology. Circulation 129(7):798–810

Sagiv M, Goldhammer E, Ben–Sira D, Amir R (2007) What maintains energy supply at peak aerobic exercise in trained and untrained older men? Gerontology 53 (6):357–361

Smith M, Hosking J, Woodward A, Witten K, MacMillan A, Field A, Baas P, Mackie H (2017) Systematic literature review of built environment effects on physical activity and active transport – an update and new findings on health equity. Int J Behav Nutr Phys Act 14(1):158

US Department of Health and Human Services (2008) Physical activity guidelines for americans. ODPHP Publ. No. U0036. http://www.health.gov/paguidelines/pdf/paguide.pdf. Accessed on 19 Oct 2018

WHO (2008) Waist circumference and waist–hip ratio. Report of a WHO Expert Consultation. Geneva, Switzerland

WHO (2009) Global health risks: mortality and burden of disease attributable to selected major risks. WHO, Geneva

Zhang K, Werner P, Sun M, Pi–Sunyer FX, Boozer CN (2003) Measurement of human daily physical activity. Obes Res 11:33–40

Printed in the United States
By Bookmasters